STP 1265

Engineering Properties of Asphalt Mixtures and the Relationship to their Performance

Gerald A. Huber and Dale S. Decker, Editors

ASTM Publication Code Number (PCN):
04-012650-08

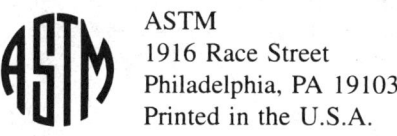

ASTM
1916 Race Street
Philadelphia, PA 19103
Printed in the U.S.A.

Library of Congress Cataloging-in-Publication Data

Engineering properties of asphalt mixtures and the relationship to
  performance / Gerald A. Huber and Dale S. Decker, editors.
      p. cm. -- (STP ; 1265)
    "ASTM publication code number (PCN) 04-012650-08."
    Includes bibliographical references and index.
    ISBN 0-8031-2002-8
    1. Asphalt emulsion mixtures--Testing. 2. Pavements, Asphalt-
  -Testing. 3. Pavements, Asphalt concrete--Testing. I. Huber,
  Gerald A. II. Decker, Dale S., 1952-   . III. Series: ASTM special
  technical publication ; 1265.
  TE275.E64 1995
  625.8'5--dc20   95-46522
                                                                   CIP

Copyright © 1995 AMERICAN SOCIETY FOR TESTING AND MATERIALS, Philadelphia, PA. All rights reserved. This material may not be reproduced or copied, in whole or in part, in any printed, mechanical, electronic, film, or other distribution and storage media, without the written consent of the publisher.

Photocopy Rights

Authorization to photocopy items for internal or personal use, or the internal or personal use of specific clients, is granted by the AMERICAN SOCIETY FOR TESTING AND MATERIALS for users registered with the Copyright Clearance Center (CCC) Transactional Reporting Service, provided that the base fee of $2.50 per copy, plus $0.50 per page is paid directly to CCC, 222 Rosewood Dr., Danvers, MA 01923; Phone: (508) 750-8400; Fax: (508) 750-4744. For those organizations that have been granted a photocopy license by CCC, a separate system of payment has been arranged. The fee code for users of the Transactional Reporting Service is 0-8031-2002-8/95 $2.50 + .50.

Peer Review Policy

Each paper published in this volume was evaluated by three peer reviewers. The authors addressed all of the reviewers' comments to the satisfaction of both the technical editor(s) and the ASTM Committee on Publications.

To make technical information available as quickly as possible, the peer-reviewed papers in this publication were prepared "camera-ready" as submitted by the authors.

The quality of the papers in this publication reflects not only the obvious efforts of the authors and the technical editor(s), but also the work of these peer reviewers. The ASTM Committee on Publications acknowledges with appreciation their dedication and contribution to time and effort on behalf of ASTM.

Printed in Philadelphia, PA
November 1995

Foreword

This publication, *Engineering Properties of Asphalt Mixtures and the Relationship to their Performance,* contains papers presented at the symposium of the same name, held in Phoenix, AZ on 6 December 1994. The symposium was sponsored by ASTM Committee D-4 on Road and Paving Materials. Gerald A. Huber of Heritage Research Group in Indianapolis, IN and Dale S. Decker of National Asphalt Pavement Association of Lanham, MD presided as chairmen is the editors and of the resulting publication.

Contents

Overview   vi

PRODUCTS OF STRATEGIC HIGHWAY RESEARCH PROGRAM

Investigation of the Relationship Between Field Performance and Laboratory Aging Properties of Asphalt Mixtures—J. E. KUEWER, C. A. BELL AND D. A. SOSOVSKE   3

Validation of SHRP A-003A Flexural Beam Fatigue Test—J. A. DEACON, A. A. TAYEBAIL, G. M. ROWE AND C. L. MONISMITH   21

Performance-Based Properties of Asphalt Concrete Mixes—R. B. LEAHY, C. L. MONISMITH AND J. R. LUNDY   37

Development of the SHRP Superpave Mixture Specification Test Method to Control Thermal Cracking Performance of Pavements—R. ROQUE, D. H. HILTUNEN, W. G. BUTTLAR AND T. FARWANA   55

The Use of Time-Temperature Superposition to Fundamentally Characterize Asphaltic Concrete Mixtures at Low Temperatures—D. R. HILTUNEN AND R. ROQUE   74

RELATING MATERIAL PROPERTIES TO PERMANENT DEFORMATION

Behavior Analysis of Asphalt Mixtures Using Triaxial Test-Determined Properties—T. F. FWA, B. H. LOW AND S. A. TAN   97

Engineering Properties of Asphalt Mixtures and their Relationship to Performance—R. L. DAVIS   111

Evaluation of Aging Characteristics of Modified Asphalt Mixtures—S.-C. HUANG, M. TIA AND B. E. RUTH   128

Analysis of a Well-Performing Desert Pavement—P. SEBAALY, V. V. THIRMARAYAPPA, J. EPPS AND L. QUILICI   146

Evaluation of Aging of Hot-Mix Asphalt Using Wave Propagation Techniques—Y. LI AND S. NAZARIAN   166

Temperature Dependent Tensile Strength of Asphalt Mixtures in Relation to Field Cracking Data—R. P. CHAPUIS AND A. GATIEN   180

Potential of Dynamic Creep to Predict Rutting—R. B. MALLICK, R. AHLRICH AND
E. R. BROWN 194

**Comprehensive Characterization of Performance-Related Properties of Asphalt
Concrete Mixtures Through Dynamic Testing**—K. M. CHUA AND M. C. ROO 213

Overview

The Asphalt Program of the Strategic Highway Research Program (SHRP) represents a landmark in asphalt technology and marks a shift in the underpinnings of asphalt binder specification and asphalt mixture design. It is the beginning of performance based specifications for asphalt materials.

Historically, asphalt mixture design has been based upon empirical properties that, when controlled, provide mix designs with a high probability of good performance. Air voids and voids in the mineral aggregate have been recognized as important mix parameters since the 1920s. The Hubbard Field method of mix design developed in the era following World War 1 used the volumetric properties of air voids, Voice in the Mineral Aggregates (VMA) and a mechanical property test, the Hubbard Field Stability Test, as the basis for design.

The Marshall and Hveem methods of mix design that are the predominant mix design methods used today each use the same approach. Both mix design methods are based on volumetric properties and an empirical mechanical property. In both methods engineering properties are not controlled directly but are controlled indirectly through the empirical properties.

Since the development of Marshall and Hveem mix designs in the 1930s and 1940s, the research community has spent considerable effort to measure fundamental engineering properties and relate these properties to performance on the roadway. Early efforts tried to measure fundamental shear properties using various shear tests and triaxial tests adopted from the geotechnical field. Asphalt mixtures were evaluated as a partially saturated or nearly saturated three-phase system analogous to partially saturated soils. The main differences between soil and asphalt are that asphalt is more viscous than water and aggregate has a larger grain size than soils.

Also, efforts were made to measure the modulus of the asphalt mixture using different methods. Stiffness measurements were made from beams, axially loaded tall cylinders, and diametrically loaded short cylinders. Each method yielded an engineering property, but each result was different, dependent upon specimen configuration and load application.

As methods to measure laboratory properties continued to advance, efforts to relate the fundamental engineering properties to roadway performance showed less progress. Mechanical behavior of the asphalt mixtures is complex and the ability to model behavior using the measured engineering properties was questionable. Uncontrolled variables in the field could not be handled with existing models. Simplifying assumptions were required to obtain predicted performance. The prediction results were severely degraded rendering the approach less than useful.

Direct measurement of engineering properties and use of engineering properties in mix design has not become common practice for several reasons. Measuring engineering properties on hot mix asphalt is difficult. More importantly, understanding that engineering properties control performance of the completed pavement has not been obvious.

Engineering properties are difficult to quantify. They change depending upon test temperature and rate of loading. Modulus values, such as resilient modulus or creep modulus, change dramatically with temperature and confining pressure, both that vary within the pavement.

Hot mix asphalt is a composite material exhibiting visco-elastic behavior of the asphalt binder and elastic behavior of the aggregate skeleton. Early attempts to measure failure properties of the aggregate skeleton used triaxial testing. The triaxial test accounted for changes in aggregate skeleton strength from changes in confining stress but did not account for visco-elastic behavior of the asphalt binder in the skeleton.

Dynamic modulus as measured by ASTM Test Method for Dynamic Modulus of Asphalt Mixtures (D 3497) was developed to measure visco-elastic behavior of the mixture. The test, which requires servo-hydraulic equipment, can measure stiffness at different loading rates. This stiffness is a combined effect of elastic behavior of the aggregate skeleton and visco-elastic behavior of the asphalt binder. Historically, no clear method has existed to combine engineering properties measured by the triaxial test or the dynamic modulus test and relate them to permanent deformation performance.

Fatigue tests were developed to measure the resistance of asphalt pavements to repeated loading. The test required special equipment and fabrication of beam specimens. Beam fatigue behavior has been most commonly used in structural thickness design. In mixture design fatigue prediction has not been commonly used.

In summary, determining which properties most control behavior of the mixture and that can best be correlated to observed performance has been elusive.

Strategic Highway Research Program

In the mid 1980s the American Association of State Highway and Transportation Organizations (AASHTO) initiated discussion regarding a highly focused research with potential for high pay back. One area identified was asphalt binders and asphalt mixtures. Under the SHRP $50 million was earmarked for asphalt research.

The objective of SHRP was to develop a specification for asphalt binders and asphalt mixtures founded on performance based properties. Performance based properties are defined as fundamental engineering properties which are directly linked to roadway performance. Previous research was the basis for determining which properties should be measured, which tests could be used to measure the properties and how the properties could be used in a performance prediction model.

In the spring of 1993, the SHRP was completed. Products of the research program include performance based properties and performance based models which were embedded into the new Superpave method of mixture design. Superpave implementation has begun starting first with Level 1 mixture design that is based on volumetric properties. Mix design Levels 2 and 3 require greater capital expense and higher levels of training. Implementation of Superpave Level 2 and Level 3 mix design will lag several years behind Level 1.

Symposium Purpose

This symposium was intended as a forum for presentation of research to measure fundamental engineering properties and relate the engineering properties to pavement performance. Since SHRP had ended 18 months before the symposium, it was expected that SHRP researchers would present results of their work and that other researchers perusing independent lines of research would present results of their activities. As a result, leaders of this Special Technical Publication (STP) are presented with a synopsis of a portion of the SHRP research considering performance based properties that can be compared to other fundamental engineering properties.

This STP is organized into two main topic areas. The first five papers deal specifically with SHRP research results. The remaining eight papers discuss other fundamental engineering properties and methods used by the researchers to relate the properties to performance.

The first three papers in the publication, "Investigation of the Relationship Between Field Performance and Laboratory Aging Properties of Asphalt Mixtures," "Validation of SHRP A-003A Flexural Beam Fatigue Test," and "Performance Based Properties of Asphalt Concrete Mixes," all originate from the SHRP Contract A-003A. These papers deal with development of test methods used in the Superpave method of mix design. The papers also discuss material properties that the authors believe could be used to predict pavement performance in fatigue and permanent deformation. The performance prediction approach as presented offers an alternative to the performance prediction approach contained within the Superpave method of mix design.

The next two papers, "Development of the SHRP Superpave Mixture Specification Test Method to Control Thermal Cracking Performance of Pavements" and "The Use of Time-Temperature Superposition to Fundamentally Characterize Asphaltic Concrete Mixture at Low Temperatures," discuss performance based properties to predict low-temperature cracking, tests used to measure the properties and models that are used to predict low temperature cracking. The two papers form a concise documentation of development of the low-temperature cracking portion of Superpave.

The following two papers, "Behavior Analysis of Asphalt Mixtures Using Triaxial Test Determined Properties" and "Engineering Properties of Asphalt Mixtures and Their Relationship to Performance," discuss engineering properties with a longer history in performance prediction literature.

The next paper, "Evaluation of Aging Characteristics of Modified Asphalt Mixtures," investigates aging behavior of modified asphalt systems as compared to unmodified asphalt systems and the relationship of aging to pavement performance.

The next two papers, "Analysis of a Well Performing Desert Pavement" and "Evaluation of Aging Characteristics of Asphalt Concrete Using Wave Propagation Techniques," approach the problem of prediction pavement performance by investigating existing pavements, then measuring properties that can be used to explain the observed performance.

The paper "Temperature Dependent Tensile Strength of Asphalt Mixtures in Relation to Field Cracking Data" investigates the tie between tensile strength of asphalt mixtures and the resistance to low temperature cracking.

The last two papers, "Potential of Dynamic Creep to Predict Rutting" and "Comprehensive Characterization of Performance Related Properties of Asphalt Concrete Mixtures Through Dynamic Testing," investigate laboratory properties that can be used to predict pavement performance in permanent deformation and fatigue cracking.

Summary

In the future the asphalt paving industry will face increased demand for improved performance and reduced risk of premature failure. One approach to reduce risk is the ability to measure performance-based properties as part of mix design process and predict in-service performance. Efforts to develop strong links between measured engineering properties and observed performance have been ongoing for the last several decades. SHRP technology has provided an advancement that remains to be implemented.

This STP contains information about both SHRP and non-SHRP approaches. It will provide the reader with part of the background behind the Superpave method of mix design including development of test methods for fatigue cracking and permanent deformation as well as an alternative method of using the SHRP tests to predict future distress. Development of Superpave tests for low-temperature cracking is discussed, as is the performance models for predicting low-temperature cracking.

Gerald A. Huber
Heritage Research Group
Indianapolis, IN; symposium
chairman and editor

Dale S. Decker
National Asphalt Pavement Association
Lanham, MD; symposium
co-chairman and editor

Products of Strategic Highway Research Program

Julie E. Kliewer,[1] Chris A. Bell,[2] and Dan A. Sosnovske[3]

INVESTIGATION OF THE RELATIONSHIP BETWEEN FIELD PERFORMANCE AND LABORATORY AGING PROPERTIES OF ASPHALT MIXTURES

REFERENCE: Kliewer, J. E., Bell, C. A., and Sosnovske, D. A., **"Investigation of the Relationship Between Field Performance and Laboratory Aging Properties of Asphalt Mixtures,"** Engineering Properties of Asphalt Mixtures and the Relationship to their Performance, ASTM STP 1265, Gerald A. Huber and Dale S. Decker, Eds., American Society for Testing and Materials, Philadelphia, 1995.

ABSTRACT: Short- and long-term aging procedures were developed at Oregon State University under the Strategic Highway Research Program (SHRP) A-003A project. Several alternative methods for short- and long-term aging of asphalt-aggregate were examined. These were subjected to extensive laboratory and field validation test programs. Laboratory validation was done with an extensive testing program using 8 asphalts and 4 aggregates. Field validation utilized 20 test sites and compared the modulus of field aged and laboratory aged specimens.

For short-term aging a procedure of curing the loose mix in a forced draft oven at 135°C for 4 hours was recommended. For long-term aging, oven aging at 85°C for 4 days was recommended to represent projects of about 10 years old. Alternative procedures may be appropriate in some situations. Long-term oven aging for 2 days at 100°C could be used for stiff mixes, whereas a low pressure procedure of 4 days at 85°C may be necessary for soft mixes.

The results of the asphalt-aggregate mixture testing presented herein show that the aging of the mixture is dependant on both the asphalt and the aggregate. Also, it appears, from the evaluation of data from other SHRP contractors, that the aging and subsequent testing of asphalt alone is not a good predictor of how a mixture will behave due to the effect of the asphalt-aggregate interaction.

Continued monitoring of field projects is needed, particularly for Dry-No Freeze and Wet-No Freeze zones. Increasing the number of sites and total number of

---

[1]Assistant Professor, Department of Forest Engineering, Oregon State University, Corvallis, OR 97331.

[2]Professor, Department of Civil Engineering, Oregon State University, Corvallis, OR 97331.

[3]Senior Research Specialist, Oregon Department of Transportation, Salem, OR 97310.

specimens tested will facilitate the use of regression analyses to develop performance prediction models.

KEYWORDS: aging, asphalt-aggregate mixtures, asphalt-aggregate interaction, field performance, laboratory testing, resilient modulus

---

INTRODUCTION

The development and validation of laboratory aging procedures to simulate short- and long-term aging for asphalt-aggregate mixtures was undertaken as part of the Strategic Highway Research Program (SHRP) project A-003A at Oregon State University (OSU). This work has been previously described by Bell et al., [1], [2], [3]. The purpose of this paper is to summarize those aspects of the work that relate to relationship between field performance and laboratory aging of asphalt mixtures.

The recommended laboratory procedure for short-term aging is to heat the loose mix in a forced draft oven for 4 hours at a temperature of 135°C. This short-term oven aging (STOA) simulates the aging of the mixture during the construction process.

The recommended laboratory procedure for long-term aging of the compacted mixture is to heat the compacted specimens in a forced draft oven at 85°C. This long-term oven aging (LTOA) simulates the aging of the mixture in service. An aging period of 2 days aging appears representative of young projects of up to 5 years old. An aging period of 4 or 5 days is recommended to simulate projects about 10 years old. Different environmental zones are accommodated by changing the aging period. Low modulus mixtures may require an alternate long-term aging method such as low pressure oxidation (LPO) by passing oxygen through the compacted specimens while they are confined in a triaxial cell [2].

Two testing programs were conducted to thoroughly evaluate the recommended methods and establish their validity. The first program, designated laboratory validation (discussed in detail by Bell and Sosnovske, [2] and, by Sosnovske et al., [4]) involved tests on laboratory prepared mixture specimens using 8 different asphalts and 4 different aggregates. The second program, designated field validation (discussed in detail by Bell, Wieder and Fellin, [3]), involved tests on laboratory prepared mixtures and field cores from 20 projects varying from 0 to 19 years old.

EXPERIMENT DESIGN

Laboratory Validation Program

Eight different asphalt types and four different aggregates were used to develop 32 mixture combinations. The materials used for this testing program were selected from those stored at the SHRP Material Reference Library (MRL). The aggregates used represent a broad range of aggregate characteristics, from a high absorption crushed limestone to a river run gravel. The asphalts used cover a broad range of sources and grades. The materials used are described in Table 1. Laboratory specimens were prepared with target air voids of 8±1 percent. Target air voids were selected to represent the high side of field conditions for dense mixtures and to match voids used in other aspects of the SHRP A-003A project.

TABLE 1--<u>Materials Used.</u>

Aggregate		Asphalt	
MRL Code	Description	MRL Code	Grade
RC	Limestone (high absorption), basic	AAA-1	150/200
RD	Limestone (low absorption), basic	AAB-1	AC-10
RH	Greywacke, acidic	AAC-1	AC-8
RJ	Conglomerate, acidic	AAD-1	AR-4000
		AAF-1	AC-20
		AAG-1	AR-4000
		AAK-1	AC-30
		AAM-1	AC-20

Four different long-term aging procedures were examined: low pressure oxidation (LPO) at 60°C and 85°C for five days, long-term oven aging (LTOA) at 85°C, all for five days, and LTOA at 100°C for two days. All long-term aged specimens were first short-term aged at 135°C for 4 hours before compaction. Only the data from STOA and LTOA at 85°C will be reported herein, since these are the methods recommended for use.

<u>Field Validation Program</u>

This program was designed to compare the performance of laboratory aged mixtures with that of field projects. Twenty projects were considered that represented "new," "young" and "old" pavements and a range of climatic regions. Projects were selected, in part, based on the availability of retained asphalt and aggregate, mix design information, and cooperation from the controlling agency for field coring. The target air voids for each laboratory specimen was the field air void content ±1 percent. The details of this program are described by Bell et al [3]. Table 2 summarizes some basic data for each site.

The four "new" sites located in Oregon were used to validate STOA. Laboratory mixtures were subjected to 0, 4 or 8 hours of STOA at 135°C prior to compaction. The modulus of compacted laboratory mixtures was compared with that obtained for compacted field produced materials.

The nine "young" projects and seven "old" projects were all subjected to four hours of STOA at 135°C. Long-term oven aging was then conducted at either 85°C for 0,2,4 or 8 days, or, at 100°C for 0, 1, 2 and 4 days. Only the data for STOA and LTOA at 85°C will be reported herein. Both sets of projects served to validate the recommended short-term aging and long-term aging methods.

TABLE 2--Summary of Field Validation Sites.

a) New Projects

Site and Project Number	Construct Date	Asphalt Grade	Asphalt Content[1] (%)	Admix	Climate
Stag Hollow - Wapato Road, #913	1990	AC-15	6.2	None	Wet-No Freeze
Butter Creek - Old Oregon Trail, #816	1990	AC-15	5.9	None	Dry-Freeze
Rock Cr - Anlauf, #852	1990	AC-20	5.3	PBS[2]	Wet-No Freeze
Lobert, #874	1990	AC-15	5.8	Lime	Dry-Freeze

b) Young Projects

Site and Project Number	Construct Date	Asphalt Grade	Asphalt Content[1] (%)	Admix	Average Air Voids (%)	Climate
Arizona SPS-5 (AZ5)	1990	AC-40	4.7	PC-II[3]	4.3	Dry-No Freeze
Arizona SPS-6 (AZ6)	1990	AC-20	4.6	Lime	4.8	Dry-Freeze
California AAMAS Batch (CAB)	1989	AR-4000	5.6	-	6.3	Dry-Freeze
California AAMAS Drum (CAD)	1989	AR-4000	4.5	-	6.0	Dry-Freeze
California GPS-6 (CAG)	1991	NA[4]	5.2	-	5.6	Dry-No Freeze
French A Section	1986	40/50 Pen	5.9	-	6.5	Dry-No Freeze
French B Section	1986	40/50 Pen	5.9	-	6.5	Dry-No Freeze
French C Section	1986	40/50 Pen	5.9	-	6.5	Dry-No Freeze
Georgia AAMAS (GAA)	1989	AC-30	4.3	Lime	7.6	Wet-No Freeze
Michigan SPS-6 (MI6)	1990	AC-10	5.6	Flyash	4.1	Wet-Freeze
Minnesota SPS-6 (MN6)	1990	NA	5.6	-	5.6	Wet-Freeze
Wisconsin AAMAS (WIA)	1989	200/300 Pen	5.3	Recycle	3.7	Wet-Freeze

c) Old Projects

Site and Project Number	Construct Date	Asphalt Grade	Asphalt Content[1] (%)	Admixture	Average Air Voids (%)	Climate
SR-14, #1801	1973	85/100 Pen	6.0	None	6.6	Wet-No Freeze
SR-522, #6048	1977	AR-4000W	5.7	None	6.0	Wet-No Freeze
SR-167, #6049	1972	85/100	5.8	None	3.2	Wet-No Freeze
SR-12, #1002	1988	AR-4000W	5.9	None	4.4	Dry-Freeze
US 97, #1006	1982	AR-4000W	5.3	PBS[2]	5.1	Dry-Freeze
US 195, #1008	1978	AR-4000W	6.0	PBS	7.1	Dry-Freeze
US 195, #6056	1986	AR-4000W	6.2	PBS	4.6	Dry-Freeze

[1]By weight of total mix
[2]Pave Bond Special
[3]Type II Portland Cement
[4]Not Available

AGING METHODS

Laboratory Validation Program

No aging--Three specimens were prepared, by compacting immediately after mixing, to represent an "unaged" condition. These specimens were prepared in the same manner as all others except that they were not short-term aged. As soon as mixing was complete, the specimens were placed in an oven and brought to the proper equiviscous temperature for that mix. The specimens were then compacted using a California Kneading Compactor. An equiviscous temperature of 665 ± 80 centistokes was adopted by the SHRP A-003A researchers for all specimens compacted using the California Kneading Compactor.

Short-Term Aging--All specimens except the unaged specimens were subjected to short-term aging. The short-term aging procedure subjected the loose mixture to a four hour curing period in a forced draft oven at 135°C prior to compaction. During the curing period, the loose mixture is spread in a pan at a rate of approximately 20 kg per square meter. The mix is stirred and turned once an hour to ensure uniform aging throughout the sample. After the short-term aging, the samples are brought to the equiviscous temperature of 665 ± 80 centistokes and compacted using a California Kneading Compactor.

Long-Term Oven Aging--Long-term oven aging was used to simulate field aging. The procedure is carried out on compacted specimens after they have been subjected to short-term aging. The specimens were placed in a forced draft oven at 85°C for five days. Alternatively, a temperature of 100°C for two days was used. After the aging period, the oven is turned off and left to cool to room temperature before removing the specimens. The specimens were not tested until at least 24 hours after removal from the oven.

Field Validation Program

The aging methods used were the same as those described for the laboratory validation project with the following exceptions:

Short-Term Aging--In addition to STOA at 135°C for 4 hours, mixtures were also subjected to 8 hours of STOA at 135°C ("New" projects only).

Long-Term Oven Aging--In this program, LTOA at 85°C was conducted for 0, 2, 4 and 8 days. For the laboratory validation program, only 5 days was used. For LTOA at 100°C, this program used 0, 1, 2 and 4 days of aging. The laboratory validation program used only 2 days.

EVALUATION METHODS

Resilient Modulus

The resilient modulus was determined at 25°C using the diametral (indirect tension) (ASTM D 4123) and the triaxial compression modes of testing with a 0.1 second load time at a frequency of 1 Hz. A constant strain level of 0.01% was maintained throughout the test.

Only the results of the resilient modulus data for the diametral mode of testing will be summarized here. The diametral modulus data showed less variability than the triaxial modulus data; approximately ± 10% compared to ± 15% for the triaxial modulus data. This difference can be attributed, in part, to the relatively short specimen used (10.2 cm) in the triaxial mode. Details of the data for both diametral and triaxial modes of testing have been presented by Bell et al. [2], [3]. For the field validation program many of the field cores were of insufficient length to enable triaxial modulus to be determined.

RESULTS

Laboratory Evaluation Program

Calculation of Modulus Ratio--To analyze the effects of short- and long-term aging on asphalt-aggregate mixtures the aging ratio was determined. To compute this ratio, a measure of the unaged modulus was needed to compare to the modulus from the aged specimens. For each mixture, three specimens were prepared without being subjected to STOA, instead, these specimens were compacted as soon as they could be brought to the proper compaction temperature. These specimens were said to be in an "unaged" condition and each specimen was tested for resilient modulus in both the diametral and the triaxial mode.

Variability of the unaged modulus was ignored in the computation of the modulus ratio and in any subsequent statistical analysis. In all but a few cases, the unaged specimens were found to have a somewhat different air void level than the STOA specimens. Typically, the unaged specimens had about one percent lower air voids than the STOA specimens. The difference in air voids resulted from a failure to adjust the compactive efforts to account for the reduced stiffness of the unaged mixture. This made it necessary to adjust the modulus values of the STOA specimens to correspond to the same air void level as the unaged specimens in order to compute a meaningful aging ratio.

To make this adjustment, the relationship between modulus and air voids was considered for the unaged specimens over the entire data set. It was found that, on the average, the modulus decreased 100 MPa for each 1 percent increase in air voids. Knowing this and the average modulus and air void level for each combination, the unaged modulus at any void level could be estimated.

Short-Term Aging Results--The diametral modulus ratios (STOA diametral modulus divided by unaged diametral modulus) by asphalt are shown for each of the four aggregates by the dark bars in Figure 1. The asphalt showing the greatest aging (in terms of diametral modulus change) has the highest ratio.

Long-Term Aging Results--The modulus ratios (long-term aged diametral modulus divided by adjusted unaged diametral modulus) for LTOA at 85°C are shown by asphalt in ranking order in Figure 1 by the light bars.

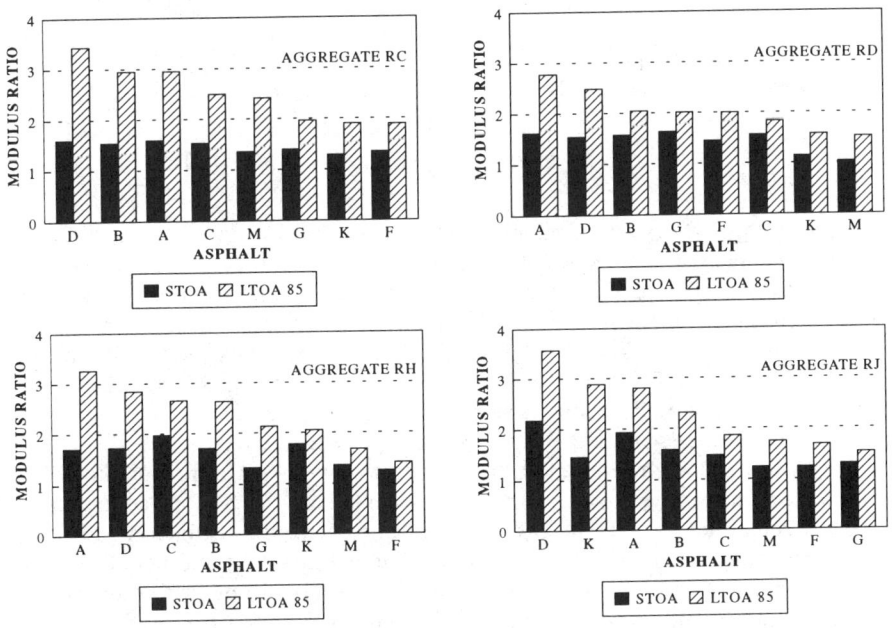

FIG. 1--Diametral Modulus Ratios for Short- and Long-Term Oven Aging

Field Validation Program

New Projects--Figure 2 shows a typical result from this phase of the field validation. The field modulus value was measured on cores taken from the in-service pavement. The intersection of the value of modulus for the field material with the laboratory data gives an estimate of the amount of STOA needed to represent short-term aging in the field. Similar data were obtained for each of the four new projects. The range of STOA times to age the laboratory specimens to equal those from the field ranged from 4.5 hours to over 12 hours. A STOA procedure of 4 hours was selected for future use. This is consistent with that recommended by Von Quintus et al. [5], and is deliberately conservative (with respect to amount of aging) because it was found that specimens became difficult to compact after longer periods of STOA.

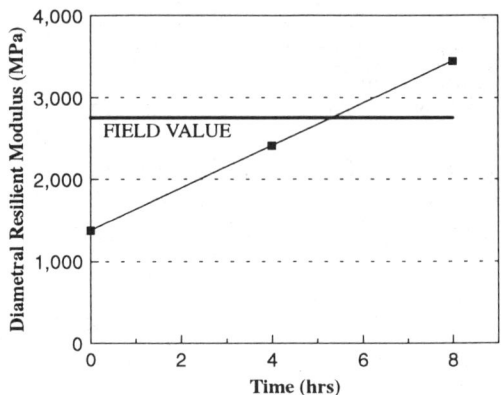

FIG. 2--Short-Term Oven Aging at 135°C - ODOT #816.

   <u>Young Projects</u>--A statistical analysis of the moduli for field cores and laboratory specimens was done to determine which of the laboratory treatments most closely matched the field aging for each site. Both Tukey and LSD (least significant difference) multiple comparison approaches were used. The LSD approach was favored, because it produces tighter confidence intervals, allowing differences to be better detected. The LSD method is commonly used when planned comparisons are made, as in this study, with several treatments being compared to the field [6]. Only examples and summaries of the LSD analyses are presented here.
   Figures 3 and 4 illustrate the LSD statistical analyses performed on the young and old projects respectively. The vertical lines plotted are the confidence intervals and the horizontal lines are the means, for each set of diametral resilient modulus values. Any aging treatment with a confidence interval that does not include any of the field confidence interval is considered to be statistically different from the field. This might result in one or more aging treatments being different from field aging; i.e. in Figure 3, the "unaged" treatment is different from the field mean, and in Figure 4, all treatments are different from the field mean. Table 3 summarizes the results for young and old sites respectively for long-term oven aging at 85°C and 100°C by listing the aging treatments with means not significantly different from the field mean.

ANALYSIS OF RESULTS

Laboratory Evaluation Program

   <u>Short-Term Aging of Asphalt-Aggregate Mixtures</u>--The data presented in Figure 1 suggests that mixture aging susceptibility is aggregate dependent. However, the effect of the asphalt is more significant. The rankings of the eight asphalts based on short-term aging (Figure 1) vary with aggregate type. Table 4a presents the rankings numerically and shows where groups of asphalts are statistically similar using Waller groupings [7]. In particular, asphalt AAK-1 moves around in the rankings,

showing relatively little aging with aggregates that are basic (RC and RD) and relatively high aging with the aggregates that are acidic (RH and RJ).

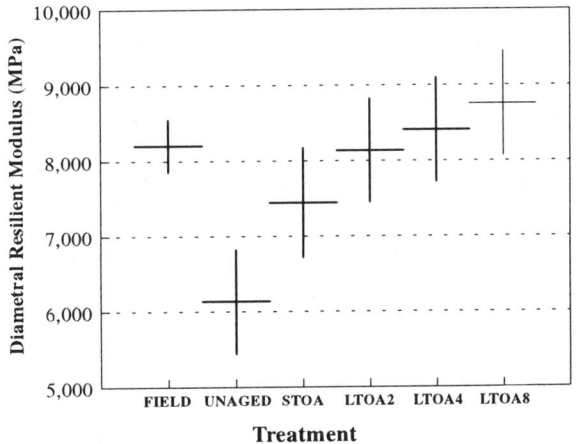

FIG. 3--95% LSD Comparison for a Young Project (AZ SPS-5 for LTOA at 85°C)

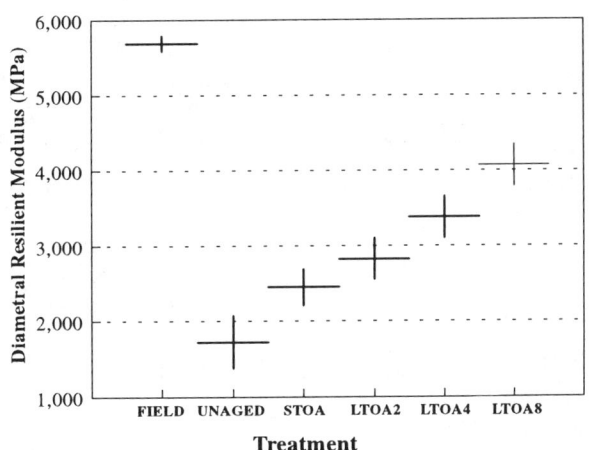

FIG. 4--95% LSD Comparison for an Old Project (WA #1008 for LTOA at 85°C)

Long-Term Aging of Asphalt-Aggregate Mixtures--The data for long-term aging (Figure 1) support those for short-term aging, i.e., they also suggest that aging is aggregate dependant, as well as, asphalt dependant. Table 4b presents the rankings numerically and shows where groups of asphalt are statistically similar, using Waller groupings[7]. Note that there appears to be more differentiation among asphalts following long-term aging than following short-term aging.

Table 3--Comparison of Aging Treatment Means and Field Means

A)   "Young" Projects--

Site	Age	Aging Treatments with Means "Not Significantly Different" from Field Mean	
		LSD (85°C)	LSD (100°C)
AZ SPS-5 (AZ5)	6 months	STOA, LTOA 2,4,8	LTOA 1, 2, 4
AZ SPS-6 (AZ6)	Few months	UNAGED, STOA	UNAGED, STOA
CA Combined (CAC)	> 2 years	UNAGED	UNAGED
CA GPS-6 (CAG)	Few months	STOA	STOA
French Combined (FRC)	5 years	LTOA 8	NONE
GA AAMAS (GAA)	2 years	NONE	NONE
MI SPS-6 (MI6)	6 months	STOA	STOA, LTOA 1, 2
MN SPS-6 (MN6)	1-1/2 years	UNAGED, STOA, LTOA 2	UNAGED, STOA, LTOA 1
WI AAMAS (WIA)	>3 years	LTOA 2, 4	LTOA 2

B)   "Old" Projects

Site	Age (yr)	Aging Treatments with Means "Not Significantly Different" from Field Mean	
		LSD (85°C)	LSD (100°C)
1801	18	LTOA 8	NONE
6048	14	STOA, LTOA 2, 4, 8	STOA, LTOA 1*, 2, 4
6049	19	LTOA 4, 8	LTOA 4
1002	3	STOA	STOA
1006	9	NONE	NOT TESTED
1008	13	NONE	NONE
6056	5	LTOA 2, 4	LTOA 1*, 2

*Not modulus tested, but assumed to fall within the field LSD interval.

Key: LSD = Least Squared Difference
STOA = Short Term Oven Aging
LTOAi = Long Term Oven Aging for i days
NONE = All of the aging treatments are significantly different from the field mean.

Comparison of Mixture Aging with Asphalt Aging--Aging of asphalt cement was investigated by the SHRP A-002A contractor. Data for original (tank), thin film oven (TFO) aged, and pressure aging vessel (PAV) aged asphalt have been presented in several A-002A reports. These routine data were summarized by Christensen and

Anderson [8]. As with mixture aging data, the asphalt aging data can be used to calculate an aging ratio based on the aged viscosity at 60°C compared to the original viscosity at 60°C. It is recognized that there may be a better measure of the aging of asphalt cement (different test and/or temperature), however, these data were not readily available at the time these rankings were prepared. The asphalts can be then ranked in order of aging susceptibility.

Table 4--Rankings of Modulus Ratio by Aggregate

A) Short-Term Oven Aging

Aggregate	more susceptible				Ranking							less susceptible
RC	D	>	B	>	C	>	A	>	G	>	M	>	F	>	K
	1.59		1.53		1.52		1.58		1.39		1.35		1.34		1.28
RD	G	>	A	>	B	>	C	>	D	>	F	>	K	>	M
	1.62		1.61		1.56		1.55		1.53		1.44		1.14		1.03
RH	C	>	K	>	D	>	B	>	A	>	M	>	G	>	F
	1.97		1.78		1.72		1.70		1.70		1.36		1.32		1.26
RJ	D	>	A	>	B	>	C	>	K	>	G	>	F	>	M
	2.18		1.93		1.58		1.47		1.45		1.30		1.24		1.24

B) Long-Term Oven Aging at 85°C.

Aggregate	more susceptible				Ranking							less susceptible
RC	D	>	B	>	A	>	C	>	M	>	G	>	K	>	F
	3.43		2.95		2.95		2.49		2.41		1.96		1.90		1.89
RD	A	>	D	>	B	>	G	>	F	>	C	>	K	>	M
	2.78		2.48		2.04		2.01		2.00		1.83		1.56		1.51
RH	A	>	D	>	C	>	B	>	G	>	K	>	M	>	F
	3.26		2.84		2.65		2.63		2.13		2.05		1.67		1.41
RJ	D	>	K	>	A	>	B	>	C	>	M	>	F	>	G
	3.58		2.88		2.80		2.31		1.86		1.73		1.67		1.52

Note: Waller groupings of statistically similar behavior are underscored. A,B,C,D,F,G,K, & M represent asphalt codes, AAA-1, etc. given in Table 1. Numerical values, eg. 1.59, are ratios of aged to un-aged diametral moduli.

Short-Term Aging--Table 5 shows rankings for mixtures based on short-term aging and the asphalt rankings based on thin film oven (TFO) aging. It should be noted that TFO aging is analogous to short-term mixture aging, and that (as with mixture rankings), the difference between some asphalts is not statistically significant. Nevertheless, it is clear that there is little relationship between the mixture rankings and the asphalt rankings. The major similarity is that asphalt AAM-1 is one of the two "best" asphalts in both the mixture and asphalt short-term aging. A major difference is that asphalt AAK-1 is ranked one of the two "worst" from asphalt TFO aging and among two of the "best' if STOA aging with aggregates RC and RD is considered.

Table 5--Comparison of Rankings for Short-Term Aging Mixtures and Asphalt Alone

	Ranking of Asphalt (by Aggregate)					
	A-003A[1]					A-002A[2]
	RC	RD	RH	RJ	A-003A Rankings	None
Most susceptible	D	G	C	D	D	D
	B	A	K	A	A	K
	C	B	D	B	C	F
	A	C	B	C	B	C
	G	D	A	K	K	B
	M	F	M	G	G	A
	F	K	G	F	F	M
Least susceptible	K	M	F	M	M	G

[1]Based on short-term aging ratios from diametral modulus.
[2]Based on data reported by Christensen and Anderson (1992).

Long-Term Aging--Table 6 shows the rankings for mixtures based on LTOA at 85°C, and, rankings for asphalt developed from the data reported by Christensen and Anderson [8]. Also summarized are rankings developed from data reported by Robertson et al., [9], for asphalt recovered from "mixtures" of single size fine aggregate and asphalt subjected to aging in the pressure aging vessel (PAV).

As with the short-term aging comparisons, there is little similarity between the rankings for long-term aging of asphalt mixtures and asphalt alone. In fact, there is even less similarity, since, asphalt AAM-1 appears to have more susceptibility to long-term aging in the PAV than it does in the TFO, (relative to the other asphalts) and has moved in the rankings.

There is more similarity between the rankings based on mixture aging and those based on the data for fine aggregate mixtures developed by the A-002A contractor. However, the rankings are different as indicated in Table 6.

General Discussion--The aggregate in an asphalt-aggregate mixture does not affect the oxidation (production of oxidative products such as carbonyl and sulfoxide) of an asphalt [10], [11]. However, aging of the mixture (as measured by stiffness) is

Table 6--Comparison of Rankings for Long-Term Aging of Mixtures and Asphalts

	Ranking of Asphalts (by Aggregate)								
	A-003A[1]				A-002A[2]	A-002A[3]		A-002A[4]	
	RC	RD	RH	RJ	None	RD	RJ	RD	A-003A Rankings
Most susceptible	D	A	A	D	D	$F_\uparrow^4$*	D	$F_\uparrow^4$*	D
	B	D	D	K	F	$M_\uparrow^6$*	B	$M_\uparrow^6$*	A
	A	B	C	A	M	D	$F_\uparrow^4$*	$C_\uparrow^3$*	B
	C	G	B	B	K	C	C	D	C
	M	F	G	C	C	$A_\downarrow^4$*	M	G	K
	G	C	K	M	B	K	$A_\downarrow^3$*	$A_\downarrow^5$*	G
	K	K	M	F	A	$G_\downarrow^3$*	$K_\downarrow^5$*	$B_\downarrow^4$*	M
Least susceptible	F	M	F	G	G	$B_\downarrow^5$*	G	K	F

[1]Based on long-term aging ratios from diametral modulus for LTOA at 85°C.
[2]Based on data for TFO-PAV aging [8].
[3]Based on PAV aging at 60°C for 144 hours. Prior short-term aging [9].
[4]Asphalt alone was subjected to TFO aging prior to mixing and PAV aging [10].
*Arrows and adjacent numbers indicate changes in ranking relative to A-003A rankings. Changes of one or two places are not shown.

affected by the aggregate. Jones [10] explains this by saying: "The viscosity of a given asphalt at a given time and temperature is a function of the molecular weight and the network that has been formed among the groups of polar molecules. How the polar networks form is a function of their environment, in this case the aggregate." The difference in rankings between mixtures and asphalt, based on either short-term or long-term aging data support this statement and indicate the need for mixture testing to evaluate the aging susceptibility of a mixture. Clearly, the aging of the asphalt alone, or in a fine aggregate mixture, is not an indicator of how a mix will age because the environment is different.

The influence of the aggregate on mixture aging appears to be related to the chemical interaction of the aggregate and the asphalt as suggested by Jones [10]. This interaction may be related to adhesion; the greater the adhesion, the greater the mitigation of aging. The mixture aging rankings given in Tables 5 and 6 suggest this hypothesis, since the rankings are similar for the two "basic" aggregates (RC and RD) and for the two "acidic" aggregates (RH and RJ). Some of the asphalts rank similarly regardless of the aggregate types, whereas others (such as AAG-1 and AAK-1) behave very differently according to the aggregate types. It is known that asphalt AAG-1 was lime treated in the refining process and it is therefore rational that it would exhibit good adhesion and reduced aging tendency with the "acidic" aggregates (RH and RJ) such as indicated by the short-term aging data. However, the rankings of asphalt AAG-1 for long-term aging do not appear to be influenced by aggregate type.

Attempts were made to predict the aging behavior of an asphalt-aggregate mixture (relative to other mixtures) by considering the acidity or basicity of the asphalt and of the aggregate. The percent acid components (weak acids plus strong acids) and the percent base components (weak base plus strong base) were determined for each asphalt using asphalt property data from the MRL. It was hypothesized that asphalts with a high percentage of acid components would age less with the basic aggregates than the acidic aggregates and the reverse would be true for asphalts with a high percentage of basic components. Although there appears to be some tendency for this to be true, this simple hypothesis could not be validated by the available data, in part, because the experiment was not designed with this in mind.

Field Validation Program

Figures 5, 6, and 7 are summaries of the statistical analysis for young and old projects. They show aging treatments which were similar to the field, and also show where the average field modulus lies relative to the average modulus for each of the aging treatments. The age of each site is also noted, and the sites are grouped according to climatic region. Since the three French sections utilized the same asphalt grade (but different suppliers) and were subjected to the same traffic and environmental conditions, the data from those sites were combined. The same was done for the California AAMAS sites, where both drum and batch plants were used. Figures 5 and 6 indicate and Table 3 shows, that for the young projects, five of the seven sites between 0-2 years old had field means that were at least statistically similar to the STOA aging treatment. Two of the five also had field moduli similar to unaged specimens. The two sites that were not similar to STOA (Georgia and California AAMAS) had field cores which came from weak or damaged pavements. This resulted in field moduli equal to or lower than the unaged specimens. This data further validates the preliminary study conclusion that 4 hours of STOA at 135°C is a good estimate of short-term aging since "young" projects exhibited similar levels of aging. The two oldest sites in this group of "young" projects, Wisconsin (3 years) and France (5 years), required short-term and long-term oven aging to match the field modulus average.

Figure 7 and Table 3 show that five of the "old" sites required at least 8 days of long-term oven aging to match the field modulus mean. Sites 1006 and 1008 (9 and 13 years old) in a Dry-Freeze portion of Washington, had field moduli that were significantly higher than any of the aging treatments. The youngest site 1002 (3 years) required only STOA to match the field, while the 5 year old site 6056 had a field modulus similar to 2 and 4 days of 85°C aging.

CONCLUSIONS AND RECOMMENDATIONS

Conclusions

The following conclusions can be drawn from the results of this study:

1) The aging of asphalt-aggregate mixtures (as measured by the change in resilient modulus) is influenced by both the asphalt and aggregate. Aging of the asphalt alone, and subsequent testing does not appear to be an

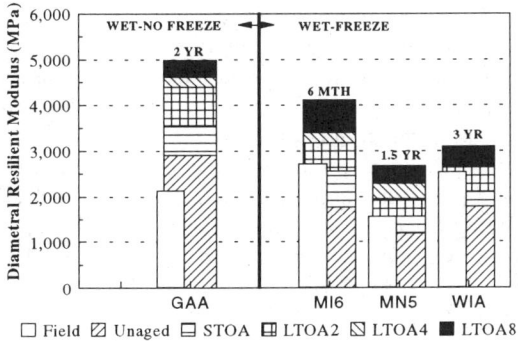

Fig. 5--Field Validation - Modulus Comparisons, Wet-Freeze/No Freeze Zones, "Young" Projects.

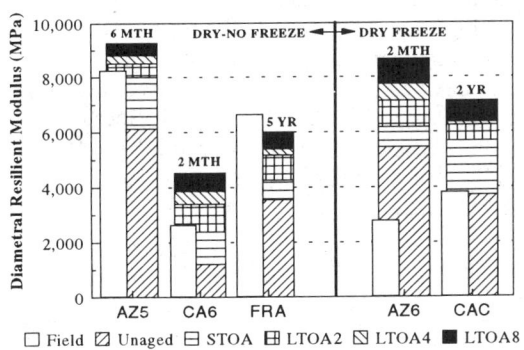

Fig. 6-- Field Validation - Modulus Comparisons, Dry-Freeze/No Freeze Zones, "Young" Projects.

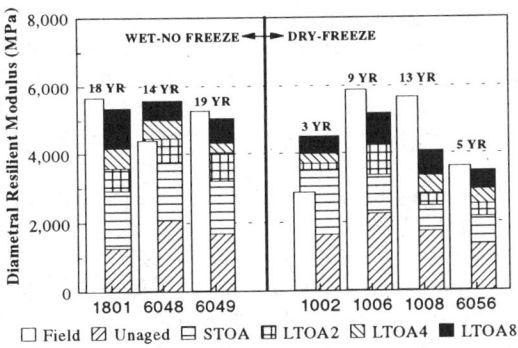

Fig. 7--Field Validation - Modulus Comparisons, Dry-Freeze/Wet-No Freeze Zones, "Old" Projects.

adequate means of predicting mixture performance, with respect to aging, because of the apparent mitigating effect aggregate has on aging.

2) The aging of certain asphalts is strongly mitigated by some aggregates, but not by others. This appears to be related to the strength of the chemical bonding (adhesion) between the asphalt and aggregate.

3) The short-term aging procedure produces a change in resilient modulus of up to a factor of two. For a particular aggregate, there is not a statistically significant difference in the aging of certain asphalts. The eight asphalts investigated typically fell into three groups, i.e., those with high, medium and low aging susceptibility.

4) Based on the study of new and young field sites, 4 hours of oven aging at 135°C appears representative of the short-term aging which occurs in the field during mixing and placement. This is also representative of young projects less than two years old.

5) Two days of long-term oven aging at 85°C is representative of pavement up to five years old, depending on the climate.

6) Four days of oven aging at 85°C appears to be representative of field aging of about 15 years in a Wet-No Freeze zone and about 7 years in a Dry-Freeze zone.

7) It was not possible to develop guidelines for Wet-Freeze or Dry-No Freeze zones, because no projects of sufficient age could be located.

8) 100°C oven aging for 1, 2, and 4 days achieves similar stiffness to 85°C aging for 2, 4, and 8 days, but may damage the samples in the process. 85°C aging is considered to be more reliable.

Recommendations

The following recommendations may be made from the results of this study:

1) Continued monitoring of field projects is needed, particularly for Dry- No Freeze and Wet-Freeze zones. Increasing the number of sites and the total number of specimens prepared will facilitate the use of regression analysis to develop prediction models. The sites selected should have in-service lives ranging from 1 to 20 or more years to encompass all long-term aging in the field. A reduction in the 95 percent confidence intervals found with the LSD analyses would improve the correlation of the laboratory procedures with the age of the field cores.

2) The field validation study addressed validation of the 4 hour at 135°C Short-Term Oven Aging and the Long-Term Oven Aging at 85°C and 100°C. One additional method for Long-Term Aging, Low Pressure Oxidation, was developed at OSU and deserves additional validation study. This approach may be necessary for mixtures with relatively low modulus. The aging effects of Low Pressure Oxidation have been evaluated in the development of alternate laboratory aging procedures by Bell and Sosnovske [2]. The pressures involved are not high enough to pose safety problems.

3) Further work is required to determine how aggregate influences aging in an asphalt-aggregate mixture using an appropriately designed experiment.
4) Aging behavior of recycled mixtures, mixtures with modified binders, and open graded mixtures should be investigated.

ACKNOWLEDGEMENTS

The work reported herein was conducted as part of project A-003A of the Strategic Highway Research Program (SHRP). SHRP is a unit of the National Research Council that was authorized by section 128 of the Surface Transportation and Uniform Relocation Assistance Act of 1987. This project is entitled, "Performance Related Testing and Measuring of Asphalt-Aggregate Interactions of Mixtures," and was conducted by the Institute of Transportation Studies, University of California, Berkeley, with Carl L. Monismith as Principal Investigator. The authors wish to acknowledge J. Claine Peterson for his thoughtful review of this work.

DISCLAIMER

The contents of this report reflect the views of the authors, who are solely responsible for the facts and accuracy of the data presented. The contents do not necessarily reflect the official view or policies of the Strategic Highway Research Program (SHRP) or SHRP's sponsors. The results reported here are not necessarily in agreement with the results of other SHRP research activities. They are reported to stimulate review and discussion within the research community. This report does not constitute a standard, specification, or regulation.

REFERENCES

[1] Bell, C.A., AbWahab, Y., Cristi, M. E., and Sosnovske, D. A., Selection of Laboratory Aging Procedures for Asphalt-Aggregate Mixtures, SHRP-A-390, Strategic Highway Research Program, Washington, DC, 1994.

[2] Bell, C.A., and D. Sosnovske, "Validation of A003A Hypothesis for Aging," TM-OSU-A003A-92-14, Final Report to Strategic Highway Research Program, Washington, DC, 1992.

[3] Bell, C.A., A. Wieder, and M.J. Fellin, Laboratory Aging of Asphalt-Aggregate Mixtures: Field Validation, SHRP-A-383, Strategic Highway Research Program, Washington, DC, 1994.

[4] Sosnovske, D.A., AbWahab, Y., and Bell, C. A., "Role of Asphalt and Aggregate in the Aging of Bituminous Mixtures," Transportation Research Record 1386, TRB, Washington, DC, 1993, pp: 10-21.

[5] Von Quintus, H.L., Scherocman, J. A., Hughes, C. S., and Kennedy, T. W., "Asphalt-Aggregate Mixture System," NCHRP Report 338, National Cooperative Highway Research Program, Washington, DC, 1991.

[6] Ramsey F., and Schafer, D., Statistical Data Analysis, Department of Statistics, Oregon State University, Corvallis, OR, 1990.

[7] Waller, R.A., and Kemp, K.E., "Computations of Bayesian t-Values for Multiple Comparisons," Journal of Statistical Computation and Simulation, 75,pp: 169-172, 1976.

[8] Christensen, D., and Anderson, D., "Interpretation of Dynamic Mechanical Analysis Test Data for Paving Grade Asphalt Cements," Journal of the Association of Asphalt Paving Technologists, Volume 61, Charleston, SC, March 1992, pp: 67-116.

[9] Robertson, R., "Quarterly Report for SHRP Contract A-002A," December 1991.

[10] Jones, D. R., "An Asphalt Primer: Understanding How the Origin and Composition of Paving-Grade Asphalt Cements Affect Their Performance," SHRP Asphalt Research Program Technical Memorandum #4, 1992.

[11] Curtis, C. W., Ensley, K., and Epps, J., Fundamental Properties of Asphalt-Aggregate Interactions Including Adhesion and Absorption, Strategic Highway Research Program, Report SHRP-A-341, National Research Council, Washington, D. C., 1993.

John A. Deacon,[1] Akhtarhusein A. Tayebali,[2] Geoffrey M. Rowe,[3] and Carl L. Monismith[4]

VALIDATION OF SHRP A-003A FLEXURAL BEAM FATIGUE TEST

REFERENCE: Deacon, J. A., Tayebali, A. A., Rowe, G. M., and Monismith, C. L., **"Validation of SHRP A-003A Flexural Beam Fatigue Test,"** Engineering Properties of Asphalt Mixtures and the Relationship to their Performance, ASTM STP 1265, Gerald A. Huber and Dale S. Decker, Eds., American Society for Testing and Materials, Philadelphia, 1995.

ABSTRACT: The recently completed Strategic Highway Research Program (SHRP) Project A-003A included the development of a flexural beam test for measuring fatigue of asphalt mixes and an analysis system based on multilayer elastic theory to estimate in-situ performance. The ability of laboratory measurements—coupled with simulations of load response in situ—to predict in-situ pavement performance is demonstrated. Specifically the current investigation shows that controlled-strain laboratory tests are suitable for pavement performance predictions; that A-003A test measurements are sensitive to a variety of fundamental mix properties known to affect fatigue performance and that they compare favorably with measurements using other state-of-the-art equipment; and that, using calibrations developed by the investigation, in-service pavement performance can be predicted for a variety of mixtures, climates, pavement structures, and traffic loadings. Attempts to simulate mixture performance under the accelerated testing of both laboratory wheel tracking and ALF experimentation yielded mixed results.

KEYWORDS: asphalt mixes, pavement performance, fatigue, mix design, validation, laboratory testing, multilayer elasticity

---

Included within the scope of the recently completed Strategic Highway Research Program (SHRP) Project A-003A was the development of an accelerated performance-related test for characterizing the fatigue response of asphalt-aggregate mixes and the

---

[1]Professor of Civil Engineering, University of Kentucky, Lexington, KY 40506.

[2]Assistant Professor of Civil Engineering, North Carolina State University, Raleigh, NC 27695.

[3]Technical Vice President, SWK Pavement Engineering, Millington, NJ 07946.

[4]Professor of Civil Engineering, University of California, Berkeley, CA 94720.

development of an analysis system for determining effects of mix properties on pavement performance. A central task was validation of this testing and analysis system to assure that engineering properties measured in the laboratory could be combined with the analysis system to yield reasonably accurate estimates of the effects of mix properties on pavement performance. This paper summarizes the validation process. Complete documentation is available elsewhere [1].

Conceptually validation of the testing and analysis system is a hierarchical process involving three orders or levels. First-order validation demonstrates that laboratory tests measure fundamental properties which are sensitive to mix composition and which conform with a priori notions about mix, loading, and temperature effects. First-order validation is accomplished largely through the test development process.

Second-order validation demonstrates that the testing and analysis system can accurately discriminate between superior and inferior mixes based on their demonstrated performance in wheel-track tests, in accelerated full-scale simulations, and/or in controlled or uncontrolled field trials. Laboratory wheel-track testing was the primary means available for second-order validation given project constraints on time and budget.

Finally, third-order validation demonstrates that the testing and analysis system yields accurate predictions of primary and secondary responses of mixes in wheel-track tests, in accelerated full-scale simulations, and/or in field trials. Although direct third-order validation within Project A-003A eventually proved to be infeasible, a preliminary linkage was established between pavement performance predictions based on American Association of State Highway and Transportation Officials (AASHTO) pavement design guidelines and estimates based on the A-003A testing and analysis system.

Addressed in subsequent sections of this paper are 1) development of the accelerated performance test for fatigue, 2) first- and second-order validation activities, and 3) linkages between in-service pavement performance and mix properties, climate, pavement structure, and traffic.

DEVELOPMENT OF ACCELERATED PERFORMANCE TEST

Development of the accelerated performance-related fatigue test was the first, and perhaps most important, step in the validation process. This section describes the testing apparatus and sketches the process, using a combination of literature review, pilot testing, and performance simulation, through which it evolved.

Candidate Test Procedures

An extensive literature review revealed five primary candidate test procedures for accelerated performance-related testing in fatigue. These candidates included pulsed third-point loading in flexure, sinusoidal cantilever loading in flexure, sinusoidal tension-compression loading uniaxially, notched beam flexural loading coupled with a fracture mechanics (C*-line integral) interpretation, and pulsed loading in the diametral mode. Evaluation focused on these five tests because each met the basic requirement for measuring fundamental material properties under accelerated laboratory loading, and each was among the most widely used and most critically evaluated of those currently available. The evaluation included both a qualitative component based principally on the literature review and a quantitative component based on laboratory testing. Mix variables in the pilot test program included asphalt type, asphalt content, aggregate type, degree of compaction, temperature, and strain level.

The evaluation highlighted many of the advantages and disadvantages of the candidate tests. Two of the candidates—uniaxial tension tests and fracture mechanics tests—were quickly eliminated after preliminary testing. Gripping the specimen is difficult in pure tension testing, and end-cap failure due to stress concentrations was a persistent problem. Testing for fracture mechanics analysis was thought to be too extensive for routine mix analysis and design: repetitive fatigue tests are necessary to evaluate the crack initiation process as well as the crack growth rate, and notched-beam strength tests are necessary to evaluate the C^*-line integral.

Among the remaining three candidates, the diametral (indirect tension) test was particularly appealing because of its ability to test briquette-shaped specimens. The pilot testing program demonstrated that, although diametral fatigue was reasonably reliable, it was generally inferior to flexural fatigue in the sensitivity of its measurements to mix composition. Measured stiffnesses were comparatively large—perhaps excessively so—and cycles to failure were unreasonably small. In the final analysis, diametral testing was judged to be unsuitable for routine mix analysis and design because of 1) the large incidence of unacceptable fracture patterns, 2) stress concentrations at the loading platens, and 3) its limitation to a controlled-stress mode of loading. Moreover, its variable biaxial stress state, its inability to reverse stress fields, and the confounding influence of permanent deformations within test specimens on their resistance to repetitive tensile loading raised serious additional concerns.

The testing program revealed no striking differences between beam and cantilever testing. However, beam measurements were convincingly more sensitive to mix variables than cantilever measurements. With the exception of beam testing's failure in the pilot testing to reasonably demonstrate the effect of asphalt content on cycles to failure and cantilever testing's questionable stiffness-temperature effects, the results of both tests were judged to be reasonable.

In summary although beam tests are advantageous because of their uniform stress distribution and the fact that gluing is unnecessary, the beam and cantilever tests are considered equivalent means for assessing the fatigue behavior of asphalt-aggregate mixes. Nevertheless, the beam test was eventually selected by Project A-003A because of the researchers' familiarity with it and because of the sophistication of its current design and its software interface.

<u>Simulation of Field Conditions</u> [2]

One key question involved mode of loading. Mode of loading is of potential importance in mix analysis because, for similar initial conditions, fatigue life is typically greater in controlled-strain loading than in controlled-stress loading, and, even more importantly, mixes of greater stiffness tend to perform better in controlled-stress loading but worse in controlled-strain loading. The critical unknown in mix evaluation was whether the mode of loading selected for laboratory testing would influence the results of the evaluation process. For example, if Mix A were to be judged superior to Mix B on the basis of laboratory tests under one mode of loading, would it also be superior on the basis of results obtained under the other mode of loading?

Answering this question is not as straightforward as it might at first appear. The fatigue behavior of an in-situ mix is determined by two key factors: 1) the mix's resistance to the destructive effects of repetitive stresses or strains and 2) the level of stress or strain to which it is subjected under traffic loading. Laboratory testing is necessary for establishing the fundamental fatigue behavior, and mechanistic analysis is necessary for establishing critical levels of stress or strain. Thus a combination of fatigue testing and mechanistic analysis is required for evaluating likely in-situ behavior. Multilayer elastic analysis was incorporated into the A-003A mix testing and analysis system because of its relative simplicity and widespread availability and because of its extensive prior use in fatigue investigations. As is common, the maximum principal tensile strain at the bottom of the asphalt layer was considered to be the primary determinant of fatigue distress.

To support an assessment of possible mode-of-loading effects, the pilot test program yielded two sources of information including 1) complete and reliable flexural stiffness and fatigue data for three mixes under both modes of loading and 2) fatigue life calibrations (as functions of air-void content, initial strain level, and initial flexural stiffness), again under both modes of loading. Two sets of simulations were performed: the first included the three tested mixes and the second included four hypothetical mixes—mixes with stiffnesses of 2 756 and 4 134 MPa (400 000 and 600 000 psi) and air-void contents of 4% and 7%—with fatigue properties defined by the regression equations. The surface thickness of the two-layered structures ranged from 5 to 30.5 cm (2 to 12 in), and subgrade stiffnesses were varied to exaggerate the changing relative stiffness of the surface layer to that of its support. Poisson's ratios of 0.35 and 0.30 were used for surface and subgrade layers, respectively. Loading consisted of 44 kN (10 000 lb) on 690-kPa (100-psi) dual tires, spaced 30.5 cm (12 in) center-to-center. Adjustments were made in both the calibrations and the simulations to assure that the same stiffness values were used for each mix under both controlled-strain and controlled-stress conditions.

Results of the first set of simulations (Table 1) show that the performance ranking of the three mixes is unaffected by mode of loading. Mix 3 was always superior for all structural sections, and Mix 1 was always inferior. Results of the second set of simulations (Table 2) generally seem to be independent of mode of loading as well. For the same mix stiffness, low air-void mixes were always superior to high air-void mixes. As anticipated initially stiffer mixes demonstrated inferior fatigue resistance for "thin" pavements but superior resistance for "thick" ones. The only differences between modes of loading were found at the borderline between "thin" and "thick" pavements—7.6 to 12.7 cm (3 to 5 in). Based on this analysis, it appears that controlled-stress and controlled-strain testing may yield similar mix rankings—especially for the substantial pavement structures characteristic of the nation's primary trucking highways—when test results are interpreted in terms of the expected pavement performance. As a result, either mode of loading appears suitable for the laboratory component of a comprehensive mix design and analysis system.

TABLE 1--Mode-of-loading effect on simulated fatigue life of tested mixes.

Mode of Loading	Surface Thickness (cm)	Subgrade Stiffness (MPa)	Simulated Fatigue Life		
			Mix 1 (3 964 MPa)	Mix 2 (4 930 MPa)	Mix 3 (6 984 MPa)
Controlled-Stress Testing	5.1	207	2 000	10 000	40 000
	10.2	172	6 000	35 000	213 000
	15.2	137	17 000	107 000	1 275 000
	20.3	103	39 000	257 000	5 524 000
	25.4	69	77 000	528 000	17 209 000
	30.5	34	129 000	957 000	40 394 000
Controlled-Strain Testing	5.1	207	5 000	11 000	27 000
	10.2	172	16 000	35 000	65 000
	15.2	137	50 000	96 000	167 000
	20.3	103	126 000	212 000	363 000
	25.4	69	269 000	408 000	660 000
	30.5	34	476 000	688 000	1 035 000

TABLE 2--Mode-of-loading effect on simulated fatigue life of hypothetical mixes.

Mode of Loading	Surface Thickness (cm)	Subgrade Stiffness (MPa)	Simulated Fatigue Life			
			4% Air-Void Content		7% Air-Void Content	
			2 756 MPa	4 134 MPa	2 756 MPa	4 134 MPa
Controlled-Stress Testing	5.1	207	46 000	30 000	25 000	16 000
	10.2	172	103 000	91 000	56 000	50 000
	15.2	137	290 000	294 000	159 000	161 000
	20.3	103	695 000	766 000	381 000	420 000
	25.4	69	1 324 000	1 641 000	725 000	899 000
	30.5	34	2 201 000	2 929 000	1 205 000	1 604 000
Controlled-Strain Testing	5.1	207	54 000	49 000	13 000	12 000
	10.2	172	137 000	174 000	33 000	42 000
	15.2	137	449 000	672 000	108 000	162 000
	20.3	103	1 225 000	2 021 000	295 000	487 000
	25.4	69	2 565 000	4 847 000	618 000	1 168 000
	30.5	34	4 599 000	9 425 000	1 109 000	2 272 000

Test Apparatus

The flexural beam (third-point loading) fatigue test was adopted for further development and use on the basis of 1) the literature review, 2) the pilot test program, and 3) the researchers' prior experience with this test method. Controlled-strain loading was selected because 1) the loading and control systems under development were particularly suitable for strain control, 2) with proper interpretation, controlled-strain and controlled-stress loading had been found to yield similar ranking of alternate mixes, and 3) controlled-strain loading was more compatible with the crack propagation concept and the performance prediction models that were being developed by SHRP Project A-005.

The test apparatus can be operated in a stand-alone configuration or as a module with the Universal Testing Machine that was developed by Project A-003A for the permanent-deformation program. Test specimens, placed horizontally in the test frame, are 6.4 cm (2.5 in) wide, 5.1 cm (2 in) high, and 38.1 cm (15 in) long. Loading is applied vertically by hydraulic load actuators, and automatic-testing-system software is used for test control and data acquisition. All testing reported herein was performed on sawed specimens. The sinusoidal loading of 0.1-second duration had a frequency of 10 Hz, and the initial specimen position defined one extreme of the range in specimen deformations under load. Because the weight of the specimen and the applied load act in different directions, the dynamic strain amplitude, computed from the change in deformation under load, includes a strain reversal. Failure of the specimen was defined to coincide with the point at which repeated loading had reduced its stiffness to 50% of the initial value.

FIRST-ORDER VALIDATION

The test development process comprised the initial phase of first-order validation. This was followed by 1) two series of tests designed to determine the sensitivity of test measurements to mix composition and to compare the reasonableness of these measurements with a priori notions and 2) further comparative testing with other flexural fatigue equipment to confirm the similarity of test measurements.

Sensitivity to Mix Properties

The sensitivity of test measurements to mix properties was investigated by two test series, an expanded (8x2) test experiment and a mix design experiment. The 8x2 experiment was a full factorial design with replicates of 32 mixes (eight asphalts, two aggregates, and two air-void contents). The mix design experiment included six additional mixes (two asphalt contents and three air-void contents). General Linear Modeling (GLM) was used to evaluate the statistical significance of mixture and strain-level effects.

The 8x2 experiment confirmed the statistical significance of all main mixture effects and several interaction effects as well (Table 3). Moreover, it generally confirmed the a priori notion that compositional changes which increase the initial stiffness of the mix are inclined to reduce its performance under controlled-strain loading. Thus, although asphalt and aggregate effects are complex, mixes generally perform poorer under controlled-strain loading when they incorporate asphalts and aggregates which increase initial mix stiffness (Table 4). As expected, increased air-void content, despite reducing initial mix stiffness, resulted in reduced fatigue life (Table 4). The larger number and size of air voids likely increase the opportunity for cracks to form as well as the rate at which they propagate: such effects apparently exceed any effects of reduced initial mix stiffness due to increased air voids.

TABLE 3--Statistically significant effects in GLM (8x2 experiment).

Factor/Interaction	Initial Flexural Stiffness	Fatigue Life
Asphalt Source	H	H
Aggregate Source	H	H
Air-Void Content	H	H
Strain	H	H
Asphalt Source * Aggregate Source	H	H
Asphalt Source * Air-Void Content	H	H
Asphalt Source * Strain		
Aggregate Source * Air-Void Content		H
Aggregate Source * Strain	B	
Air-Void Content * Strain		

NOTES:

Description	Probability of type 1 error
H = highly significant	less than 0.01
S = significant	0.01+ to 0.05
B = barely significant	0.05+ to 0.10
Blank = not significant	greater than 0.10

TABLE 4--Average fatigue properties (8x2 experiment).

Effect	Initial Flexural Stiffness (MPa)	Fatigue Life (cycles)
Asphalt Source		
AAA-1 (150/200 pen)	2 037	99 300
AAB-1 (AC-10)	2 826	70 300
AAC-1 (AC-8)	3 811	41 200
AAD-1 (AR-4000)	2 663	74 400
AAF-1 (AC-20)	7 122	25 100
AAG-1 (AR-4000)	8 085	7 200
AAK-1 (AC-30)	4 087	46 200
AAM-1 (AC-20)	4 170	71 200
% Difference[a]	75%	-93%
Aggregate Source		
RH (Greywacke river gravel, partially crushed)	3 316	53 700
RD (Limestone, low-absorption crushed quarry rock)	4 666	35 100
% Difference	29%	-35%
Air-Void Content		
4%	4 404	49 400
7%	3 513	38 100
% Difference	-20%	-23%

[a]Percentage difference between AAA-1 and AAG-1

NOTES: Air-void contents adjusted to 4 and 7%.
Averages based on mean of log-transformed data.
Percentage difference is expressed as a percentage of the larger value.

The mix design experiment confirmed the statistical significance of both asphalt content and air-void content and found no significant interaction between them (Table 5). It confirmed the effect of air voids that had been observed in the 8x2 experiment and showed that, within the range investigated, increases in asphalt content were beneficial to fatigue performance (Table 6). The benefits of increased asphalt content are thought to be due to a combination of reduced initial mix stiffness and increased thickness (and, hence, reduced strains) in the asphalt coatings.

TABLE 5--Statistically significant effects in GLM (mix design experiment).

Factor/Interaction	Initial Flexural Stiffness	Fatigue Life
Asphalt Content	S	H
Air-Void Content	H	H
Strain		H
Asphalt Content * Air-Void Content	S	
Asphalt Content * Strain		
Air-Void Content * Strain		

NOTES:
	Description	Probability of type 1 error
	H = highly significant	less than 0.01
	S = significant	0.01+ to 0.05
	B = barely significant	0.05+ to 0.10
	Blank = not significant	greater than 0.10

TABLE 6--Average fatigue properties (mix design experiment).

Effect	Initial Flexural Stiffness (MPa)	Fatigue Life (cycles)
Asphalt Content		
4.5%	6 601	9 100
6.0%	6 105	27 900
% Difference	-8%	67%
Air-Void Content		
4.0%	7 762	21 500
8.0%	5 192	11 900
% Difference	-33%	-45%
Strain Level		
200 μm/m	6 350	126 200
700 μm/m	6 350	2 000
% Difference	0%	-98%

NOTES: Air-void contents adjusted to 4 and 8%.
Averages based on mean of log-transformed data.
Percentage difference is expressed as a percentage of the larger value.

This testing demonstrated that the A-003A flexural fatigue measurements are sensitive to important mix characteristics, and it confirmed findings of prior investigations regarding mix-stiffness, asphalt-content, and air-void-content effects.

Compatibility with Existing Tests

The final phase of first-order validation sought to confirm the attractiveness of the A-003A apparatus in comparison to other state-of-the-art test equipment both in terms of consistency and variability of results.

Consistency of results--Four mixtures, supplied by the Laboratoire Central des Ponts et Chaussées (LCPC), were tested not only in the University of California, Berkeley (UCB) laboratory but also in three other laboratories worldwide (SHELL-KSLA, LCPC, and SWK). Equipment included a controlled-stress, flexural-beam apparatus operated at 40 Hz (KSLA), a controlled-strain, trapezoidal-cantilever apparatus operated at 25 Hz (LCPC), and a controlled-stress, trapezoidal-cantilever apparatus operated at 20 Hz (SWK). The four mixes, differing primarily in asphalt type, asphalt content, and air-void content, included two 60/70 penetration asphalts (a more temperature-susceptible Asphalt A and a more structured Asphalt B) and a 10/20 penetration asphalt.

Results of the testing were interpreted in terms of the estimated performance of these mixes in five hypothetical pavement structures under dual tires spaced 30.5 cm (12 in) center-to-center, inflated to a contact pressure of 690 kPa (100 psi), and loaded to 22 kN (5000 lb) each. Two of the structures incorporated 10.2-cm (4-in) asphalt surfaces, one with a 30.5-cm (12-in) granular base and the other with a 43.2-cm (17-in) granular base. The remaining three structures incorporated 20.4-cm (8-in) asphalt surfaces, one with no base, one with a 15.2-cm (6-in) granular base, and the third with a 30.5-cm (12-in) base. According to AASHTO procedures, these structures are generally sufficient to accommodate design traffic ranging from 1 to 16 million ESALs [3]. Stiffnesses of the granular base and the subgrade were 207 MPa (30 000 psi) and 51.7 MPa (7 500 psi), respectively. Poisson's ratios for the surface, base, and subgrade layers were 0.35, 0.40, and 0.45, respectively. Multilayered elastic analysis was used to simulate the maximum principle tensile strain, and laboratory measurements were used to estimate fatigue life based on this strain. All testing was performed at 20°C.

Fatigue-life estimates based on A-003A testing ranked the four mixes identically based on their simulated behavior in each of the five pavement structures. Estimates from each of the other three laboratories ranked the four mixtures identically for the 10.2-cm structures as well as for the 20.4-cm structures, but the 10.2-cm rankings were somewhat different from the 20.4-cm rankings. Samples of results are shown in Fig. 1 (20.4-cm surface over 15.2-cm base) and Fig. 2 (10.2-cm surface over 30.5-cm base). While there are obvious differences among laboratories as might be expected in an experiment of this sort, the UCB (A-003A) tests seem to represent the composite of the mixture rankings quite adequately. The high-modulus mix was consistently the best performer. Mixes containing Asphalt A were generally inferior to those containing Asphalt B. The larger asphalt content in the Asphalt A mixes (5.4% compared to 4.5%) reduced initial mix stiffness and, as expected, generally prolonged fatigue life.

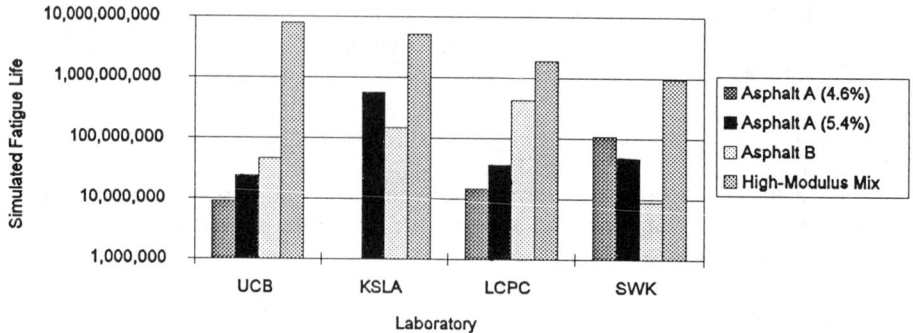

FIG. 1--Performance simulation for a 20.4-cm pavement.

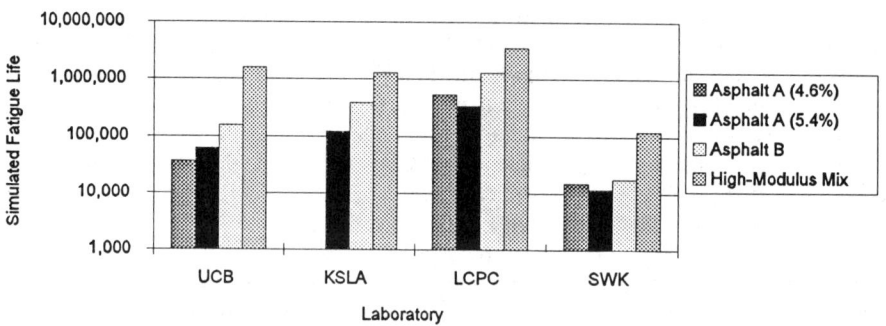

FIG. 2--Performance simulation for a 10.2-cm pavement.

Based on the above analysis, the A-003A equipment appears to be at least as reliable as other state-of-the-art equipment in its ability to rank asphalt mixes based on their simulated performance in situ.

Variability--The A-003A equipment evolved from similar third-point loading equipment that had been used in UCB laboratories for many years and that had been used in the pilot test program as well. Equipment and procedural enhancements reduced the overall testing time by a factor of approximately six and improved the repeatability of the fatigue test. Most notably, the coefficient of variation for fatigue life was reduced from about 90% with the old equipment (in the pilot testing) to about 40% with the new. This reduction is most likely due to improvements in strain control as well as to the use of larger and more uniform beam specimens.

SECOND-ORDER VALIDATION

As perceived herein, second-order validation requires more elaborate experimentation including wheel-track tests, accelerated full-scale simulations, and/or controlled or uncontrolled field trials. In support of the validation process, laboratory wheel-track testing was performed at SWK Engineering (Nottingham, United Kingdom) and at the LCPC (Nantes, France). In addition, limited accelerated full-scale simulations were performed at the accelerated loading facility (ALF) of the Federal Highway Administration (FHWA) in McLean, Virginia.

The wheel-track tests were conducted in general accord with standard practices set by the individual testing agencies. As a result there were key differences in the compaction of test slabs, the manner in which they were tested including the frequency of loading, the definition of failure, and even in the way critical strain levels were computed. The ALF testing was particularly different from the others: it was conducted in a "field" setting which included natural rainfall and temperature conditions. Comparison of the results of laboratory testing and analysis with the results of these wheel-track tests was undertaken with the full expectation that such variances among the experiments would confound the analysis and reduce its effectiveness. Field trials were within neither the scope nor the resources of the project.

Laboratory Wheel-Track Testing

The most extensive wheel-track testing, conducted by SWK, included two experiments, one evaluating SHRP Materials Reference Library (MRL) mixtures (six mixtures differing only in type of asphalt) and the second evaluating mixtures with modified asphalt binders (a control mix and three mixes with modified binders). Test slabs, measuring 1 000 mm (39.4 in) in length, 500 mm (19.7 in) in width, and 50 mm (1.97 in) in height, rested on a weak, 92-mm (3.6-in) rubber sheet overlaying a steel support. The modulus of the rubber sheet was approximately 10 MPa (1 450 psi). The magnitude of the 650-kPa (94-psi) load was adjusted to target a particular strain level in the test slab, and the load was cycled at 30 passes per minute. The test temperature was 20°C.

The primary performance measure in the wheel-track tests was the fatigue life (cycles to crack initiation) normalized to a level of 200 microstrain. Fatigue tests, performed at UCB, provided information for estimating the comparable life based on laboratory testing. The laboratory estimates were not as well correlated with the wheel-track measurements as had been expected: the squared correlation coefficient was only 0.43 (Fig. 3). Nevertheless, the four top-performing mixtures in the wheel-track testing were also top performers based on laboratory estimates. In addition, the four mixtures containing aggregate RB, ranked at the bottom of the scale from laboratory test data, also performed poorly in the wheel-track test.

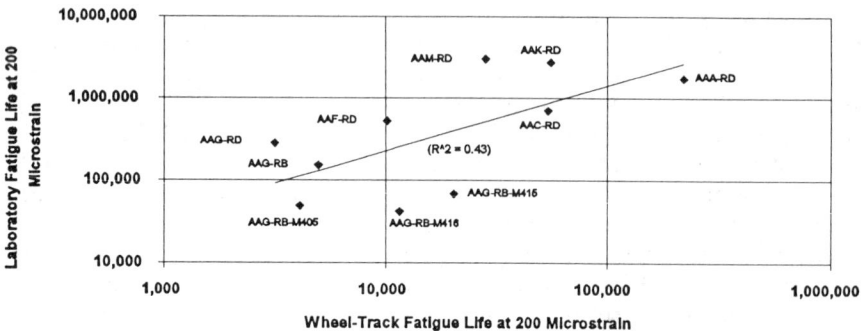

FIG. 3--Laboratory estimates of fatigue life vs. wheel-track measurements.

The accuracy of this comparison is limited by the small sample size, by differences between laboratory and wheel-track definitions of failure, by differences in compaction procedures, and by the difference in average air-void contents (averaging approximately three percent greater in the wheel-track slabs than in the laboratory beams). Moreover, the weak rubber base in the wheel-track test suggests a mode of loading more like a controlled-stress condition than the controlled-strain testing of the laboratory. Estimating wheel-track performance based on laboratory stiffness and fatigue data and on multilayer elastic analysis would likely have provided an improved comparison between laboratory and wheel track. It was not possible, however, because the magnitude of the wheel-track loading was not recorded and preserved.

The second wheel-track experiment was conducted by the LCPC. Four mixtures, previously described, were loaded in a full-scale circular track by 63.6-kN (14 300-lb) dual tires inflated to 800 kPa (116 psi) and moving at 63 km/h (39 mph). The observed performance ranking for the four mixtures was remarkably different from that simulated using multilayer elastic theory and UCB test data (Table 7). The superiority of the low asphalt-content Asphalt A mix and the inferiority of the Asphalt B mix in the wheel-track testing were both unexpected and surprising. In an effort to improve the correlation between simulated and observed performance, simulations were performed using laboratory test data from the other three laboratories (SHELL-KSLA, LCPC, and SWK), and the wheel-track experiment was repeated in its entirety. These efforts to improve the correlation were unsuccessful. Although the simulation and the wheel-track testing differed in the extent to which crack propagation was included, no satisfactory explanation for the discrepancy between simulation and observation was found.

TABLE 7--Simulated vs. observed performance in LCPC test track.

Performance Rank	Simulated Performance	Observed Performance
1 (Best)	High modulus	Asphalt A (4.6%)
2	Asphalt B	High modulus
3	Asphalt A (5.4%)	Asphalt A (5.4%)
4 (Worst)	Asphalt A (4.6%)	Asphalt B

Accelerated Full-Scale Simulation

During this investigation, the FHWA evaluated the effects of dual-tire versus single-tire loading on the fatigue cracking of an asphalt mix in an accelerated full-scale simulation (the ALF pavement test facility). Specimens from these slabs were tested using the A-003A fatigue test apparatus, and test results, together with multilayer elastic analysis, were used to simulate relative mix performance under the two loading conditions.

Simulated fatigue life was approximately 40,000 to 50,000 load repetitions for the single-tire configuration and 70,000 to 90,000 load repetitions for the dual-tire configuration. Preliminary field results indicated fatigue lives to surface crack initiation of approximately 55,000 and 110,000 load repetitions for the single- and dual-tire configurations, respectively. Not only do these simulated and measured fatigue lives compare quite favorably with each other but the simulated effect of dual tires also mirrors quite nicely the observed effect. Although this experiment is of limited use for second-order validation because it included only one asphalt mix, the similarities between simulated and measured fatigue lives lend credence to the A-003A testing and analysis system.

Summary

The wheel-track experiments were of limited worth in demonstrating that the A-003A flexural beam fatigue test can (or can not) accurately rank asphalt mixtures in terms of their fatigue performance in service. Although results in the SWK testing are favorable, additional work is clearly needed that minimizes the effect of different means for defining failure, that provides a mechanism for considering crack propagation, that allows the simulation of wheel-track performance using appropriate elastic analysis, and that reduces the level of uncertainty by increasing sample size and assuring that beam and slab specimens are as nearly alike as possible.

PAVEMENT PERFORMANCE

Performance-based mix design seeks to assure, within tolerable limits of risk, that the pavement will perform as anticipated in service. To accomplish this objective, the A-003A mix design system simulates in-service pavement performance given mix properties, climate, pavement structure, and traffic loading. The design process is an iterative one which can entail adjustments to the pavement structure as well as revisions to the mix design including the modification of binders.

To determine whether a particular mix/structure combination is satisfactory, the laboratory fatigue life is required to exceed the traffic loading (expressed in ESALs). The laboratory testing temperature at which fatigue life is measured is the critical temperature anticipated in service at the design site. The critical temperature is that temperature at the underside of the asphalt surface layer at which fatigue damage accumulates most rapidly [4]. The laboratory fatigue life is evaluated at the in-situ strain level induced by an 80.1-kN (18 000-lb) single axle load (estimated using multilayer elastic analysis). A reliability multiplier, reflecting variabilities both in laboratory measurements and in traffic estimates, is applied to the design ESALs to assure an acceptable level of risk. An empirical shift factor is also applied to the design ESALs to reflect the considerable differences between laboratory and field conditions such as the extent of crack progression, traffic wander, mix healing and aging, construction variability, etc.

Preliminary calibration of the A-003A performance model has been completed. Although future refinements are necessary, much of the calibration work, including the critical-temperature determination and the development of suitable reliability multipliers, was relatively straightforward. Reliable shift factors, on the other hand, ultimately demand field trials carefully monitored over an extended period of time. In the absence of such trials, first-order approximations of the shift factors were produced using standard AASHTO design relationships [3].

Analysis included 44 of the mixes tested by A-003A investigators, five different pavement structures (previously identified), and two regions of the country (northeast and southwest). The purpose of the analysis was to develop shift factors such that the majority of the 44 mixes, which were typical of the range in mixes currently in use in the United States, would be judged by the A-003A mix design procedure to be acceptable.

A shift factor of 13 was found to produce reasonable results at least for performance at a reliability level of 90%. Although the bulk of the mixes was acceptable at the smallest traffic loading (1 million ESALs), the percentage of suitable mixes decreased with increases in traffic loading. Although this seems to indicate that AASHTO design procedures are more conservative (vis-à-vis fatigue cracking) at smaller traffic levels than at larger ones, it also suggests that requirements for mix quality increase with increases in traffic level (despite the AASHTO requirements for thicker pavement sections with increased loading). The analysis also suggested that the loading environment for fatigue is more severe in the southwestern United States than in the Northeast. Although such differences might be reduced or possibly even eliminated by region-specific structural designs, it is not unreasonable to expect that mixes suitable for one region of the country might not be suitable for another.

Considering the fact that the A-003A mixes intentionally spanned a wide range of likely mix performance, a shift factor of 13 is certainly acceptable. However, a factor of 10 is more discriminating and is recommended initially for design applications which limit fatigue cracking to about 10% of the surface area within the wheel paths.

SUMMARY AND CONCLUSIONS

SHRP Project A-003A included the development of a flexural beam test for measuring fatigue of asphalt mixes and an analysis system based on multilayer elastic theory to estimate in-situ performance. This paper summarizes the process by which this testing and analysis system was validated. Emphasis was placed on first- and second-order validation 1) to demonstrate that the laboratory test measures fundamental properties which are sensitive to mix composition and which conform with a priori notions about mix, loading, and temperature effects and 2) to demonstrate that the testing and analysis system can accurately discriminate between superior and inferior mixes.

The validation process included a comprehensive literature review, pilot testing for evaluating primary equipment alternatives, analysis of the effects of laboratory mode of loading on estimates of pavement performance, an expanded test program for evaluating a range of mixes typical of those being used within the United States, limited comparison testing with state-of-the-art equipment being used in other countries, wheel-track tests at SWK and the LCPC, and an accelerated full-scale simulation (ALF). Although field trials were not performed, preliminary calibration of the testing and analysis system was accomplished through the use of performance relationships incorporated within the AASHTO flexible pavement design process.

Specifically the study found that controlled-strain laboratory tests provide a suitable basis for pavement performance predictions; that A-003A test measurements are sensitive to a variety of fundamental mix properties known to affect fatigue performance and that they compare favorably with measurements using other state-of-the-art equipment; and that, using calibrations developed by the investigation, in-service pavement performance can be predicted for a variety of mixtures, climates, pavement structures, and traffic loadings. Attempts to simulate mixture performance under the accelerated testing of both laboratory wheel tracking and ALF experimentation yielded mixed results.

REFERENCES

[1] Asphalt Research Program, University of California, Berkeley, "Fatigue Response of Asphalt-Aggregate Mixes," Report SHRP-A-404, Strategic Highway Research Program, National Research Council, Washington, DC, 1994.

[2] Tayebali, A. A., Deacon, J. A., Coplantz, J. S., Harvey, J. T., and Monismith, C. L., "Mixture and Mode-of-Loading Effects on Fatigue Response of Asphalt Aggregate Mixtures," Proceedings, Association of Asphalt Paving Technologists, 1994.

[3] American Association of State Highway and Transportation Officials, AASHTO Guide for Design of Pavement Structures--1986, American Association of State Highway and Transportation Officials, Washington, DC, 1986.

[4] Deacon, J. A., Coplantz, J. S., Tayebali, A. A., and Monismith, C. L., "Temperature Considerations in Asphalt-Aggregate Mixture Analysis and Design," Research Record 1454, Transportation Research Board, Washington, DC, 1994, pp. 97-112.

ACKNOWLEDGEMENTS

The research reported herein was conducted as part of Project A-003A of the Strategic Highway Research Program (SHRP). This project, entitled "Performance Related Testing and Measuring of Asphalt-Aggregate Interactions and Mixtures," was conducted principally by the Institute of Transportation Studies of the University of California, Berkeley. SHRP was a unit of the National Research Council that was authorized by Section 128 of the Surface Transportation and Uniform Relocation Assistance Act of 1987.

DISCLAIMER

This paper presents the views of the authors only and is not necessarily reflective of the views of the National Research Council, SHRP, or SHRP's sponsor. The results reported here are not necessarily in agreement with the results of other SHRP research activities. They are reported to stimulate review and discussion within the research community.

Rita B. Leahy[1], Carl L. Monismith[2], and James R. Lundy[1]

PERFORMANCE-BASED PROPERTIES OF ASPHALT CONCRETE MIXES

REFERENCE: Leahy, R. B., Monismith, C. L., Lundy, J. R., **"Performance-Based Properties of Asphalt Concrete Mixes,"** Engineering Properties of Asphalt Mixtures and the Relationship to their Performance, ASTM STP 1265, Gerald A. Huber and Dale S. Decker, Eds., American Society for Testing and Materials, Philadelphia, 1995.

ABSTRACT: Key objectives of the Strategic Highway Research Project (SHRP) A-003A were to validate properties selected for inclusion in the binder specification and to develop accelerated performance tests (APTs) suitable for predicting pavement performance in terms of fatigue and low temperature cracking as well as permanent deformation. This paper specifically addresses the laboratory tests, measured engineering properties and their relationship to the SHRP binder specification and to performance for both unmodified and modified mixes.

Materials considered in this research included 8 to 16 binders, 5 modifiers, and 2 to 4 aggregates. For fatigue, mixes were tested in the controlled-strain mode of loading using a flexural beam test device. For permanent deformation both a wheel-tracking device and a simple shear repeated-load test apparatus was used. Low temperature thermal cracking was evaluated using the thermal stress restrained specimen test (TSRST).

Overall, the results indicate that the relationship between binder properties and mix performance, as measured by the recently developed APTs, is excellent for thermal cracking, but substantially less definitive for fatigue cracking and permanent deformation. Furthermore, the recently developed APTs and resulting engineering properties are suitable for predicting pavement performance of modified mixes as well.

KEYWORDS: Asphalt-aggregate mix, fatigue, permanent deformation, thermal cracking, validation, asphalt, modified binder, binder specification

INTRODUCTION

Key objectives of the Strategic Highway Research Program (SHRP) Project A-003A were to validate properties selected for inclusion in the binder specification and to develop accelerated performance tests (APTs) suitable for predicting pavement performance in terms of fatigue and low

---

[1]Assistant Professor, Department of Civil Engineering, Oregon State University, Corvallis, OR 97331-2302.

[2]Professor of Civil Engineering, University of California, Berkeley, CA 94720.

temperature (thermal) cracking as well as permanent deformation. In addition, conditioning procedures that simulate aging and mix behavior in the presence of water (water sensitivity) were developed. The results of this major research effort are detailed in References 1 through 14.

It is the purpose of this paper to discuss the validation of the binder tests/properties developed by the A-002A contractor as they relate to the performance of asphalt-aggregate mixes, in particular to fatigue, permanent deformation, and low temperature cracking. In addition, the applicability of the mix tests for fatigue and permanent deformation to evaluate mixes containing modified binders is briefly discussed.

VALIDATION STUDY

All materials used in the validation effort were obtained from the SHRP Materials Reference Library (MRL). Eight to sixteen asphalt binders were employed for the various studies, the properties of which are reported elsewhere [15]. The asphalts selected are representative of materials currently used in the U.S. and produced from crude sources around the world. Two to four aggregates were used in the various studies. Two aggregates were employed for fatigue, permanent deformation, and thermal cracking studies. For fatigue and thermal cracking, aggregate characteristics are less significant than the asphalt properties. For permanent deformation, time and material constraints precluded the testing of more than two aggregate in spite of the universally recognized effect that aggregate has on mix resistance to rutting. Four aggregates were used for the aging and water sensitivity studies because of the dominant effect of the aggregate. These latter two studies are not reported herein. The MRL binders and aggregates used in the validation effort are shown in Table 1.

TABLE 1--Asphalt binders and aggregates used in validation effort.

Asphalts MRL Code	Grade	Aggregates MRL Code	Characteristics
AAA-1	150/200 Pen Grade	RC	Limestone, high absorption
AAB-1	AC-10		
AAC-1	AC-8 (Redwater/Gulf)		
AAD-1	AR-4000	RD	Limestone, low absorption, fully crushed quarry rock
AAF-1	AC-20		
AAG-1	AR-4000		
AAK-1	AC-30		
AAL-1	150/200 Pen Grade	RH	Greywacke, partially crushed river gravel
AAM-1	AC-20		
AAV	AC-5		
AAW	AC-20	RJ	Conglomerate, gravel
AAX	AC-20		
AAZ	AC-20		
ABA	AC-20		
ABC	AC-20		
ABD	AR-4000		

FATIGUE

Validation of the binder properties was accomplished by means of a flexural fatigue test developed as a part of the SHRP A-003A project. In addition an assessment was made of the influence of binder properties on pavement performance by computing the fatigue lives of two hypothetical pavements and comparing the estimated fatigue response with the binder properties. Both studies are described in this section.

Laboratory Flexural Fatigue Tests

For fatigue, a combination of 8 asphalts and 2 aggregates were tested in the flexural beam test device developed at the University of California at Berkeley [2]. All tests were conducted on prismatic specimens (5 cm × 6.25 cm × 37.5 cm) in the controlled-strain mode at 20°C using a sinusoidal loading at a frequency of 10 Hz.

All asphalt-aggregate mixes were prepared at a fixed asphalt content near the optimum determined by the Caltrans mix design procedure (ASTM-D1560, 1561). Mixes were prepared by rolling wheel compaction to produce specimens with target air void contents of four and seven percent.

A full factorial experiment was designed to allow all main effects and two-factor interactions to be tested. The factorial matrix consisted of 8 asphalts, 2 aggregates, 2 air void levels, and 2 strain levels resulting in a total of 64 cells. Each cell had two replicates to allow for estimation of experimental error, resulting in a total of 128 flexural fatigue tests. The factorial experiment is summarized below:

Factor	Levels
Asphalt Source	AAA, AAB, AAC, AAD, AAF, AAG, AAK, AAM
Aggregate Source	RD, RH
Air Voids	4%, 7% (target levels)
Strain Level	400, 700 μmm/mm
Replicates	2/cell
Total No. of Tests	128

Response variables included the following: a) initial flexural stiffness measured at the 50th load cycle; b) fatigue life in terms of the number of load cycles corresponding to a 50 percent reduction in flexural stiffness; and total dissipated energy, i.e., the summation of dissipated energy per cycle until a 50 percent reduction in flexural stiffness.

Binder properties provided by the A-002A contractor included complex shear modulus (G^*), phase angle (δ), storage modulus (G', which is equal to $G^*\cos\delta$), loss modulus (G'', which is equal to $G^*\sin\delta$), and loss tangent (tan δ, which is equal to G''/G'). More detailed information on asphalt binder tests and properties is presented in reference 16.

The SHRP binder specification contains the parameter $G^*\sin\delta$ obtained from a dynamic shear test at 10 rad/sec and at a temperature dependent on the asphalt grade after PAV (long-term) aging. While the mixes which tested in fatigue were subjected only to short-term aging (4 hours at 135°C), it has been shown [16] that there is an excellent correlation between the binder properties after PAV aging and after Rolling Thin Film Oven Test (RTFOT) aging which, presumably, reflects the short-term aging in the binder. Thus, if asphalt binder properties resulting from long-term binder aging in the PAV accurately represent those in mixes subjected to long-term aging, then the conclusions reported herein will probably hold for mixes subjected to long-term aging (except for possible asphalt-aggregate interaction effects on aging [10]).

Typcial results of fatigue life versus $G^*\sin\delta$ are shown in Fig. 1; it will be observed that an inverse relationship exists between binder stiffness and mix fatigue life: as binder stiffness increases, fatigue life decreases. Though not shown herein, the comprehensive statistical analysis revealed that relationships between binder properties ($G^*\sin\delta$, G^*, G') and flexural stiffness and were very strong. The relationships with dissipated energy were slightly weaker [4].

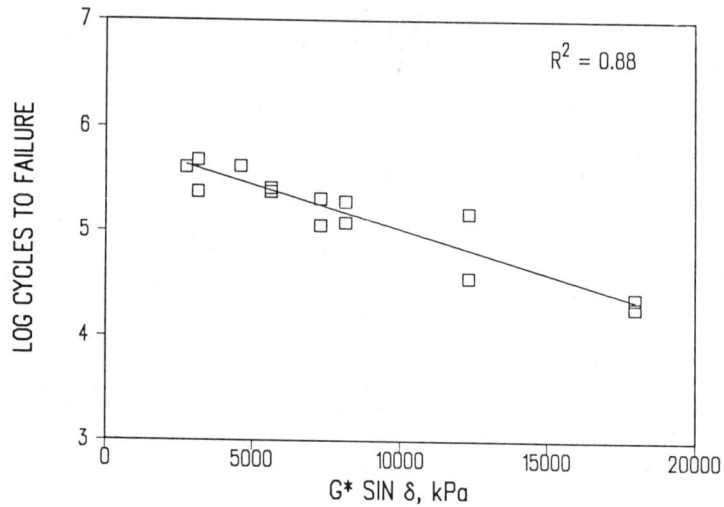

FIG. 1--Relationship between G*sinδ and log cycles to failure.

Pavement Analyses

Asphalt binder properties were compared to fatigue life estimates for "hypothetical" pavements constructed with various asphalts. These estimates were made for two hypothetical structural sections by calculating the maximum principal tensile strain (using ELSYM5) at the bottom of the asphalt concrete layer, and then calculating the corresponding fatigue life from the tensile strain using the relationship between fatigue life and strain for a given mix. The strain calculated by ELSYM5 for the hypothetical pavement was entered into the equation shown below and the corresponding fatigue life was predicted.

$$N_f = K_1 (1/\epsilon)^{K_2} \tag{1}$$

where: N_f = fatigue life,
         ϵ = strain (inch/inch), and
         K_1 and K_2 are regression coefficients.

In general, the relationship between G*sinδ and predicted pavement fatigue life was much weaker than that observed with the laboratory testing as linear regression between G*sinδ and predicted pavement fatigue life produced coefficients of determination (R^2) ranging from 0.21 to 0.38. More importantly, however, was the fact that the *direction* of the trend is opposite to that observed in the laboratory flexural fatigue analysis: in this analysis *predicted fatigue life generally increased as binder stiffness increased*.

A comparison between the SHRP binder specification for G*sinδ related to fatigue cracking and the pavement fatigue life predicted from layered elastic theory is difficult because of the reversed relationship between G*sinδ and predicted fatigue life. If confirmed in future studies that the direction of this relationship holds for certain pavement structures, the binder specification limit will need to be modified. The results of this study still indicate asphalt binder properties are important in evaluating fatigue cracking. However, the importance of considering the influence of pavement structure effects is also demonstrated.

Summary [4]

In summary, the test results and analyses suggest the following:

1. $G*\sin\delta$, as well as $G*$ and G', all result in relationships of equivalent strength with mix fatigue response. Hence, one may conclude that the effect of the "$\sin\delta$" term of $G*\sin\delta$ is negligible, and any of these terms could be used in the SHRP binder specification. However, the effect of $\sin\delta$ may still be important for modified asphalts.

2. The relationships between the binder specification property, $G*\sin\delta$, and mix flexural stiffness and fatigue life were very strong. The relationship with dissipated energy was slightly weaker.

3. In the prediction of fatigue cracking in pavement structures it appears that asphalt binder properties are again important but pavement structure effects may be equally or more important. In fact, pavement structure effects may influence fatigue cracking to the extent that the relationship between $G*\sin\delta$ and pavement fatigue life may be completely reversed as the thickness of the asphalt concrete layer changes. It is recognized that the study performed by A-003A to evaluate these effects has some limitations. Nevertheless, it identifies an issue that is worthy of further evaluation. If, after further evaluation, it is confirmed that the direction of the relationship between $G*\sin\delta$ and pavement fatigue life is dependent on the pavement structure, the binder specification will need to include provisions for pavement structure effects.

4. Overall, asphalt binder properties play a critical role in the fatigue response of asphalt-aggregate mixes. However, other mix characteristics such as air void levels and aggregate characteristics can also have a significant impact on fatigue response. Therefore, asphalt binder properties alone may not provide sufficiently reliable estimates of fatigue cracking in pavements. In critical design situations (unusual traffic volume or loading conditions, modified materials) asphalt-aggregate mix fatigue testing should be conducted to increase the reliability of estimates of fatigue cracking in pavements.

PERMANENT DEFORMATION

The relationship between binder properties and permanent deformation response of asphalt-aggregate mixes was evaluated using the wheel tracking device at the University of Nottingham and a shear device developed at the University of California at Berkeley as part of the SHRP sponsored research. Emphasis herein will be placed on the wheel tracking results.

Wheel-Track Testing

In this study, a wheel-tracking device was used to simulate the stress conditions caused by a dynamic wheel load on the pavement surface. A full factorial experiment was designed to allow all main factors and two-factor interactions to be tested. The factorial matrix consisted of 16 asphalts, two aggregates, and two air void levels, resulting in a total of 64 cells. All mixes were prepared at a fixed asphalt content near the optimum determined by the Caltrans mix design procedure (ASTM D1560, 1561). Mixes were compacted by the rolling wheel compaction method to produce specimens with target air void contents of four and seven percent. The factorial experiment is summarized below:

Factor	Levels
Asphalt Source	AAA, AAB, AAC, AAD, AAF, AAG, AAK, AAL, AAM, AAV, AAW, AAX, AAZ, ABA, ABC, ABD
Aggregate Source	RD, RH
Air Voids	4%, 7% (target levels)
Replicates	1/cell
Total No. of Tests	64

Response variables included the normalized rutting rate (mm/MPa/hr — linear regressed rut rate between 2 000 and 4 000 passes divided by contact stress) and total rut depth (mm — rut depth after 5 000 passes).[3]

The SHRP binder specification requires the value of $G^*/\sin\delta$ for any original binder to exceed 2.2 kPa when tested at 10 radians per second at the specified temperature after having been aged according to the RTFOT. The implication is that asphalt binders with $G^*/\sin\delta$ values exceeding this limit should provide acceptable resistance to permanent deformation in asphalt-aggregate mixes whereas binders with lower $G^*/\sin\delta$ values may contribute to rutting.

Asphalt binders and asphalt-aggregate mixes used in this study were subjected to similar aging and testing conditions. Asphalt binders were aged according to the RTFOT to simulate the short-term aging effects of the construction process. Asphalt-aggregate mixes were also subjected to short-term aging: after mixing, they were placed in an oven at 135°C for four hours. Asphalt binder properties were calculated for, and mixes were tested at, a temperature of 40°C. Binder properties were calculated at a loading frequency of 20 radians per second or 3.2 Hz. Considering that binder properties are logarithmic functions of loading time, the difference in loading rates is not substantial.

Wheel-tracking tests were performed by SWK Pavement Engineering Ltd. at the University of Nottingham. A wheel, fitted with a solid rubber tire, passes over the top of a 200 mm diameter cylindrical core specimen at a frequency of approximately 3 Hz or 20 radians/second. Wheel-track tests were conducted at a temperature of 40°C and each test was run for a duration of 5 000 load passes (approximately two hours). Tests were performed with an applied load producing a contact stress of approximately 730 kPa. [16]

Since it was hypothesized that asphalt source would significantly affect rutting response, analysis of variance (ANOVA) was performed to determine the influence of the various factors.

The complete statistical analyses are reported in reference 4. Typical results are shown in Figs. 2 and 3.

The results indicate that a poor relationship exists between the binder property, $G^*/\sin\delta$, and mix rutting. As engineering logic would suggest, the value of $G^*/\sin\delta$ increases as rut rate and rut depth decrease. Comprehensive statistical analysis indicated that none of the binder properties (G^*, G', or G'') were highly correlated with rut depth or rut rate. Also, there is substantial scatter in the data which suggests it would be difficult to predict rutting based solely on the binder properties. As illustrated by Figs. 2 and 3 only 18 percent to 30 percent of the variation in rutting response is explained by the parameter $G^*/\sin\delta$. Thus, most of the variation in rutting response is probably due to other variables such as aggregate characteristics and/or the testing process.

The results of this study suggest that $G^*/\sin\delta$ is not a reliable predictor of potential rutting. Aggregate and air void characteristics appear to have more influence on the rutting response of asphalt-

---

[3] SWK staff considered rut rate to be a more reliable indicator of permanent deformation performance because it is less likely to be affected by "initial start-up errors" and, perhaps, additional compaction of the specimen during the initial stages of the test.

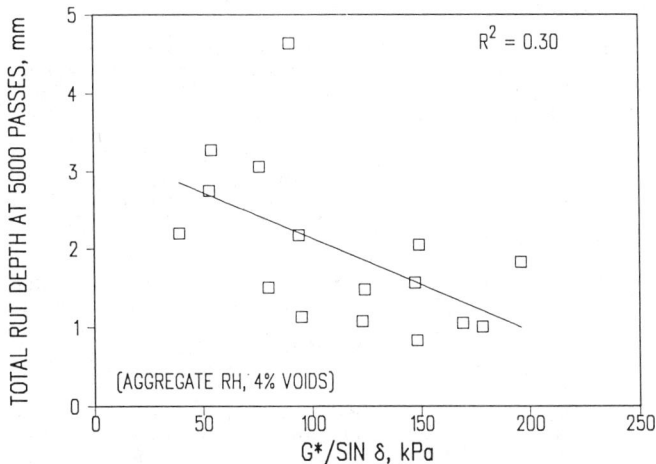

FIG. 2--Relationship between $G^*/\sin\delta$ and total rut depth.

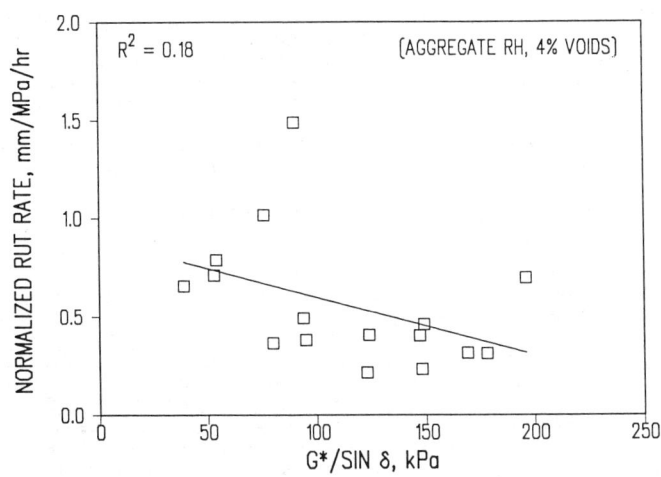

FIG. 3--Relationship between $G^*/\sin\delta$ and normalized rut rate.

aggregate mixes than does the asphalt binder. However, there are some considerations which temper this conclusion including (1) repeatability of the wheel-track test, and (2) test temperature of 40°C which may not be sufficiently high to allow the viscous characteristics of binders to affect mix rutting response. Thus the value of $G^*/\sin\delta$ may have a greater effect of rutting response than observed in this study.

Simple Shear Testing

In this portion of the investigation binder properties were compared to the permanent deformation response of asphalt-aggregate mix specimens subjected to repetitive simple shear loading under controlled conditions in the laboratory. The basis for the use of the simple shear test is contained in Reference [3].

Specimen conditioning, compaction and target void contents were as reported in the wheel track validation effort. All shear testing was conducted on cylindrical specimens 152 mm in diameter by 51 mm in height. Mixes evaluated included nine asphalts (AAB, AAC, AAD, AAG, AAK, AAM, AAV, AAZ, ABC) and the same two aggregates, and air void levels used in the wheel-tracking tests.

Half of the 72 specimens in this study were tested under a *constant height* condition (CH) and the other half were tested under a *field state of stress* (FS) condition. The CH shear test is sensitive to elastic and viscous characteristics of the asphalt binder, and it also measures the effect of dilatancy. Dilatancy in this case is the tendency of a mix to change in volume as aggregate particles are forced to slide past each other during shear deformation. The FS shear test incorporated loading conditions thought to represent the state of stress occurring in an asphalt concrete layer near the edge of a truck tire.

The CH shear test applied a cyclic (haversine) shear stress of 103 kPa ± 10% to the specimens. The load pulse duration was 0.1 second with 0.6 second between load pulses. In addition, vertical compressive loads were applied as necessary to maintain the original specimen height throughout the test duration. The magnitude of the vertical compressive load is a function of the specimen's propensity to dilate under shear loading. Shear strain was calculated from the difference between displacements measured by two LVDTs located ± 1.27 cm on each side of the at mid-height of the specimen. Each test was scheduled to run for 3 600 load cycles; all but three of the FS tests completed the scheduled 3 600 load cycles. All tests were conducted at a temperature of 60°C (140°F).

Results of an ANOVA on the shear test data indicated that asphalt source, aggregate source, and air void level each significantly affect shear response and that the influence of aggregate in the CH shear test was less than that observed in the FS shear test.

Summary

Overall, the results of this study indicate that binder properties can affect the shear response of asphalt-aggregate mixes. However, aggregate characteristics can be equally or more significant.

The results of A-003A's efforts to validate the effect of A-002A's asphalt binder properties on the permanent deformation response of asphalt-aggregate mixes indicate that the influence of asphalt is highly dependent on the conditions to which the mix is subjected. ANOVA showed the effect of asphalt was significant but that its influence was small compared to the influence of aggregate and air voids, especially when the mix was tested at lower temperatures (e.g. 40°C) or was subjected to states of stress which amplified the aggregate influence (e.g. FS shear test).

The correlations between $G^*/\sin\delta$ and the various measures of permanent deformation response were generally poor. The weak correlations are partly the result of the dominant effect of aggregate characteristics on permanent deformation response. However, in cases where mix characteristics are such that interparticle friction is low (e.g. RH aggregate and 7% air voids) and the mix is subjected to harsh environmental and loading conditions (e.g. 60°C and CH shear test), the influence of the binder becomes more readily apparent. When aggregate characteristics and/or compaction conditions are expected to result in a mix that is susceptible to permanent deformation, selection of an asphalt which can overcome these deficiencies will be important. It appears that the value of $G^*/\sin\delta$ may be used to screen binders that will provide inferior performance in such cases. The results of these studies underscore the importance of mix testing, in addition to binder testing, for evaluation of permanent deformation in pavements.

LOW TEMPERATURE (THERMAL) CRACKING

The SHRP binder specification contains a series of requirements to minimize low temperature cracking, all of which are based on tests performed on the residue of the binder from the PAV test. One of the requirements is a limiting stiffness of 300 MPa at a time of loading of 60 sec in the bending beam rheometer for a low pavement temperature corresponding to the site at which the binder will be used.

The experiment design for this task was developed to relate the properties of binders recommended by the A-002A contractor (which appear now in modified form in the SHRP binder specification) to the low temperature cracking characteristics of asphalt concrete mixes as measured by the thermal stress restrained specimen test (TSRST) (6).

The experiment design included fourteen asphalt cements and two aggregates. Two degrees of aging and two levels of air voids content were employed. A 14 × 2 × 2 × 2 × 2 replicated full factorial design was developed as follows:

Experiment Design Variable	Levels
Asphalt Type	14
Aggregate Type	2
Degree of Aging	2 (Short, Long)
Air Voids Content	2 (4%, 8%)
Rate of Cooling	1 (10°C/hr)
Replicates	2
No. of Tests	224

The MRL asphalts and aggregates used in the study are shown below.

Asphalts: AAA, AAB, AAC, AAD, AAF, AAG, AAK, AAL, AAM, AAV, AAW, AAX, AAZ, ABC
Aggregates: RC, RH

As noted above, two aging levels were considered. After mixing, the loose mix was subjected to short-term oven aging (STOA) for four hours at 135°C. Following short-term oven aging the mix was compacted. Some of the specimens were also long-term oven aged (LTOA) for five days at 85°C. Prismatic specimens (5 cm × 5 cm × 25 cm) were prepared using a Cox kneading compactor. The TSRST was used to evaluate all mixes (STOA and LTOA). Placed in a stand to ensure proper alignment, specimens were glued to end platens with an epoxy compound. After the epoxy had cured, the test specimen was cooled to a temperature of 5°C for one hour to establish thermal equilibrium prior to testing. The specimen and end platens were then placed in an environmental cabinet and cooled at a rate of 10°C/hr until fracture.

From the TSRST test results four parameters were identified to relate the fundamental properties of asphalt cement and aggregate to thermal cracking characteristics of asphalt concrete mixes. These included fracture temperature, fracture strength, slope of the thermally induced stress curve, and transition temperature. Only fracture temperature is discussed herein. Of the 224 specimens prepared, 201 were used in analysis; 23 were deemed unacceptable because void contents were not within the acceptable range.

Fracture temperature is defined as the temperature at which fracture occurs and corresponds to the temperature at which the thermal stress induced is maximum. A summary of the fracture temperature data is shown in Table 2. Mean values and the coefficients of variation of

TABLE 2--Summary statistics for fracture temperature.

Aggregate Type	Degree of Aging	Warmest Fracture Temperature (°C)	Coldest Fracture Temperature (°C)	Range (Warm-Cold)
RC	STOA	-18.6	-32.1	15.4
	LTOA	-13.6	-27.8	12.9
	Difference (STOA-LTOA)	Minimum -0.6	Maximum -6.5	Average -3.8
RH	STOA	-16.3	-32.2	15.7
	LTOA	-13.6	-29.3	14.8
	Difference (STOA-LTOA)	Minimum -0.6	Maximum -5.5	Average -2.9
Difference in STOA, °C (RC-RH)		Maximum: Minimum: Average:	-3.8 0.9 -1.16	
Difference in LTOA, °C (RC-RH)		Maximum: Minimum: Average:	-2.0 1.6 -0.42	

fracture temperature for a specific asphalt type, aggregate type, and degree of aging may be found in reference [7].

The repeatability of the TSRST for fracture temperature is quite good as the coefficients of variation for fracture temperature are typically less than or equal to 10 percent. As expected, fracture temperature varies with asphalt type. For mixes with RC aggregate, fracture temperature ranged from -32.1°C to -18.6°C and from -27.8°C to -13.6°C for STOA and LTOA aged specimens, respectively. For specimens with the RH aggregate, fracture temperatures ranged from -32.2°C to -16.3°C and from -29.3°C to -13.6°C for STOA and LTOA specimens, respectively.

Although comprehensive statistical analyses were performed to assess the influence of asphalt type, aggregate type, and degree of aging on the TSRST results, the discussion herein focuses on one aspect of the binder specification properties — the temperature corresponding to the limiting stiffness. Initially the A-002A contractor had recommended a limiting value of 200 MPa (versus the current value of 300 MPa) and the analysis presented herein is based on that requirement.

Figure 4 shows the relationship between fracture temperature in the TSRST and limiting stiffness (S(t) = 200 MPa at 2 hours) of the binder as measured by the A-002A contractor. It will be noted that an excellent correlation is obtained.

In general the results of this study suggest the following:[7]

1. The repeatability of the TSRST is estimated as good for fracture temperature.

2. Asphalt type, aggregate type, degree of aging, and air voids content are major factors which have a substantial affect on the low temperature cracking characteristics of asphalt concrete mixes. Interactions between mix properties are considered to have a minor effect.

3. Asphalt type, degree of aging, air void content, and the interaction between asphalt and degree of aging are significant factors for the fracture temperature. Fracture temperature was warmer for long-term aged mixes. Fracture temperature is most affected by asphalt type

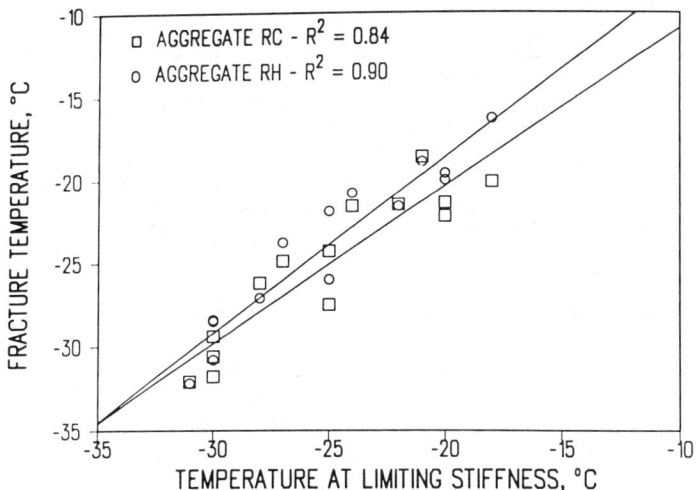

FIG. 4--Relationship between limiting stiffness and fracture temperature.

and degree of aging; also by air void content and the interaction between asphalt type and degree of aging, though to a much lesser extent.

4. Fracture temperature was highly correlated to A-002A low temperature index test results, specifically the temperature at limiting stiffness, the m-value, and the ultimate strain at failure (although only the data for temperature of limiting stiffness were reported herein).

MODIFIED MIX EVALUATION

A limited study was conducted of the influence of modified binders on mix performance in flexural fatigue and permanent deformation. The study included three MRL asphalts (AAD, AAG, and AAK), two aggregates (RB and RL), and four modifiers (coded M401, M412, M415, and M416).[4]

To establish a baseline performance level, a control (i.e. unmodified) mix was prepared for each combination of binder and aggregate. Performance comparisons were made among the various modified mixes as well as with the control mix.

Specimens for the study were fabricated by the A-004 contractor, Southwestern Laboratories (SWL). For permanent deformation evaluations the Texas-modified gyratory compactor was used to compact cylindrical specimens 15 cm in diameter and 15 cm high. For fatigue testing, block specimens 7.6 cm × 7.6 cm × 40.7 cm were compacted by SWL using a kneading compactor. The compacted specimens were shipped to UCB where they were sawed to the appropriate sizes for testing, 5 cm high for the cylindrical specimens for permanent deformation evaluation and 6.4 cm × 5 cm × 40.7 cm for the flexural fatigue specimens.

---

[4]The modifiers included SBS, SBR, EVA, reclaimed rubber, extenders, oxidants, antioxidants, mineral fillers and antistripping agents.

Permanent Deformation

Test results are presented for mixes containing combinations of asphalts AAG and AAK, aggregate RB, and modifiers M401, M412, M415, and M416. Air void contents for the specimens tested are summarized in Table 3. Two procedures were used to determine void content, one with Parafilm and one without Parafilm. Air void content determinations were made by both the A-004 and A-003A staff. UCB void measurements made without the Parafilm are similar to those made by SWL while those made with Parafilm are significantly higher. The difference is most likely because at high void contents and without Parafilm, water flows freely in and out of the specimen, and air void content typically is underestimated. Only void contents determined with Parafilm are shown in Table 3.

TABLE 3--Air void content of cylindrical specimens.

	A-004	A-003A with Parafilm
15 cm × 15 cm		
Mean (27 specimens)	6.8	12.1
Standard Deviation	0.5	1.7
Coefficient of Variation, percent	7.5	14.2
5 cm × 15 cm		
Mean (46 specimens)		11.4
Standard Deviation		2.3
Coefficient of Variation, percent		20.0

Constant height-repeated load simple shear tests were performed at 40°C. Each specimen was conditioned with 100 repetitions of a haversine shear stress of 6.9 kPa (0.1 sec haversine pulse with 0.6 sec rest period between pulses). After conditioning, the specimen was subjected to a repetitive shear stress of 70 kPa. Each test was terminated at 5 000 load cycles or a maximum shear strain of 5 percent, whichever occurred first.

Extrapolated estimates of the relative performance of the mixes over a range of void contents are shown in Figs. 5a and 5b. The average load cycles to 4 percent strain (Fig. 6) suggest the following: mixes containing binder AAG appeared to be more sensitive to modifiers than did mixes containing AAK. The performance ranking of modifiers, from best to worst, for mixes containing AAG is 401, 412, 416, 415. All of the modifiers except 415 performed better than did the unmodified control mix.

Mixes containing binder AAK were less sensitive to the influence of the modifiers. None of the modifiers improved performance. The performance ranking of the modifiers, from best to worst, for mixes containing AAK is 415, 412, 401, 416.

When the measured permanent strain was taken directly from the plots of permanent shear strain versus number of load cycles (at both 100 and 1 000 load cycles) the performance ranking of the modifiers are somewhat different (Fig. 7). For binders AAG and AAK the performance ranking, from best to worst, is 401, 412, 415, 416. This performance ranking of the modifiers, despite the difference in void contents, is nearly identical to that hypothesized by the expert task group whose members were instrumental in the selection of the modifiers to be used.

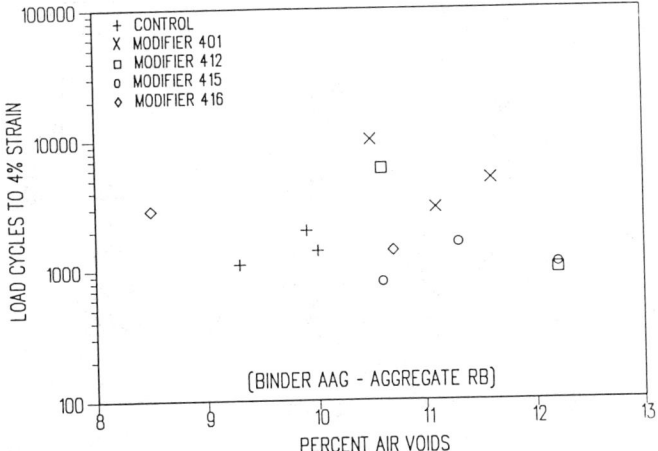

a) Modified binder AAG

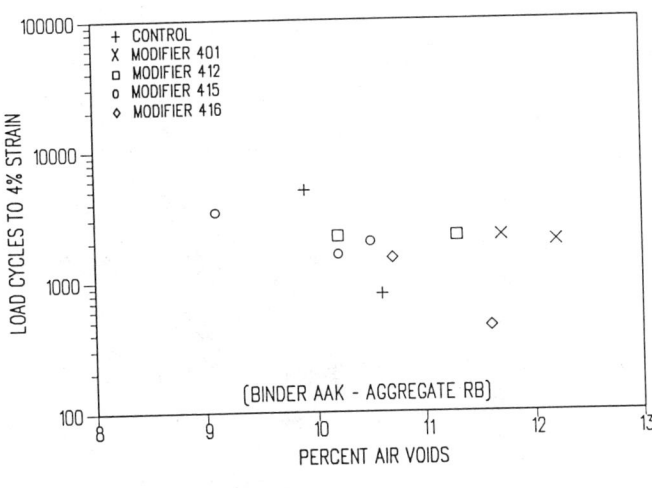

b) Modified binder AAK

FIG. 5--Performance of modified mixes over a range of air void contents.

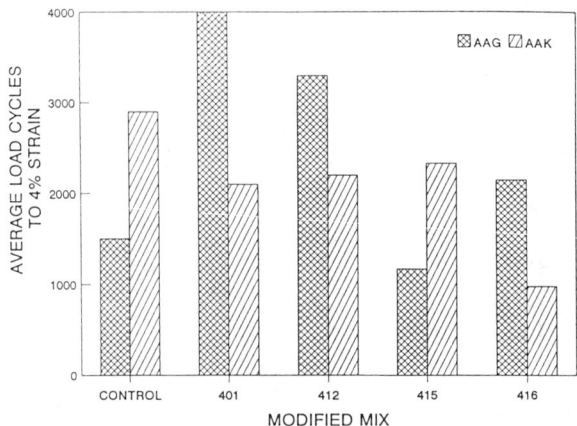

FIG. 6--Performance ranking of mixes containing modified binders — based on average cycles to 4 percent strain.

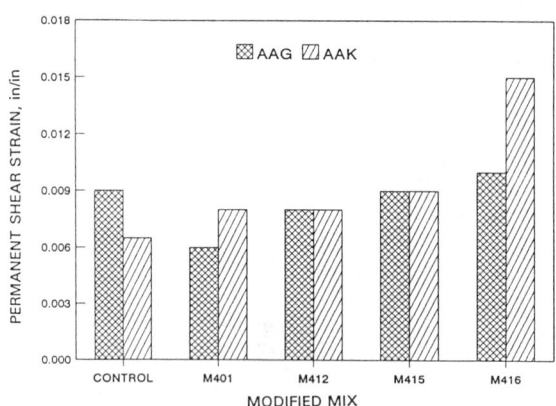

FIG. 7--Performance ranking of mixes containing modified binders, based on measured values of permanent shear strain at N=100 and 1,000 load cycles.

Fatigue

To assess the influence of modifiers on the fatigue characteristics of mixes, controlled-strain flexural fatigue tests were performed at 20°C. A sinusoidal load was applied at a frequency of 10 Hz. Three asphalts (AAF, AAG, and AAK), one aggregate (RB), and three modifiers

(M405, M415, and M416) were included in the experiment, the design for which is shown in Table 4. Response variables included initial flexural stiffness measured at the 50th load cycle; fatigue life, defined by the number of cycles to a 50-percent reduction in stiffness; cumulative dissipated energy associated with fatigue life.

TABLE 4--Features of modified asphalt mix experiment.

Number of asphalts	3 — MRL core asphalt AAF-1, AAG-1, and AAK-1
Number of aggregates	1 — MRL aggregate RB
Asphalt content	1 — 5.0 percent and 5.2 percent for unmodified and modified mixes, by weight of aggregates
Number of modifiers	3 — Modifiers identified as M405, M415, and M416
Air void levels	1 — 7±1 percent
Strain levels	2 — 10.2 and 17.8 μ m/m
Replicates at each strain level	2
Temperature	1 — 20°C (68°F)
Frequency	1 — 10 Hz (sinusoidal)
Specimen size	6 cm height, 6.25 cm width, 37.5 cm length
Method of compaction	Kneading compaction
Total number of mixes tested	10
Total number of specimens tested	39

Table 5 contains a summary of the data for all of the mixes tested. The control mixes, in order if decreasing stiffness are AAG, AAF, AAK. It should be noted that the void contents for the mixes shown in this table are those determined by the A-004 contractor without Parafilm. It is likely that the actual void contents are higher than those shown as noted earlier.

TABLE 5--Average values for stiffness, fatigue life (at 12.70 μ m/m), and cumulative dissipated energy from modified asphalt mix experiment.

Mix Type	Stiffness (MPa)	Fatigue Life (N_f) (at 12.7 μ m/m)	Cumulative Dissipated Energy to N_f (MPa)	Voids (%)	VFB[5] (%)
AAF-A	4 009	9 600	2.40	7.0	62
AAF-M405	4 700	5 200	80.90	6.6	64
AAG-1	5 413	4 300	1.58	6.4	64
AAG-M405	4 383	1 700	0.34	6.6	64
AAG-M415	4 183	1 400	0.41	6.4	65
AAG-M416	4 902	2 400	0.90	6.6	64
AAK-1	2 870	13 700	3.03	6.8	63
AAK-M405	3 263	5 800	0.90	7.1	63
AAK-M415	1 818	40 800	6.41	7.3	62
AAK-M416	1 752	69 200	10.50	7.0	63

[5]VFB = Voids filled with bitumen

52   ENGINEERING PROPERTIES OF ASPHALT MIXTURES

From the data presented in Table 5, a number of observations can be made. The fatigue life of mixes containing AAK was observed to be the longest, followed by mixes containing binders AAF and AAG. The conventional notion that stiffer mixes under controlled-strain testing performed poorer than their less stiff counterparts was confirmed for the unmodified mixes. The modified mixes, however, did not follow this pattern. Modifier M405 had a detrimental effect on all mixes, regardless of binder type. Addition of this modifier to binder AAF and AAK increase stiffness but decreased fatigue life. Modifiers M415 and M416 produced changes in performance similar to those of modifier M405 on mixes containing AAG. Addition of both modifiers reduced fatigue life, although mix stiffness decreased. For mixes containing AAK, the addition of modifiers M415 and M416 increased fatigue life substantially. In both cases, mix stiffness decreased. The modifier effects on cumulative dissipated energy were similar to those observed for fatigue life.

Though limited in extent, the results of this study indicate that both binder type and modifier type substantially affect stiffness, fatigue life, and cumulative dissipated energy. For thick sections, the ETG ranking of the modifiers in order of decreasing fatigue life was M405, M415, and M416. Results from the controlled strain testing ranked the modifiers in exactly the opposite order.

SUMMARY

The results presented in this paper suggest that the tests developed in the A-003A contract to measure the fatigue, permanent deformation, thermal (low temperature) cracking, and aging characteristics of asphalt/binder-aggregate mixes provide "reasonable" measures of mix performance.

Results of the validation study using the TSRST indicate that the SHRP binder specification requirements for low temperature cracking provide a functional measure of performance for conventional binders. Unfortunately, time did not permit performance of the TSRST on mixes containing modified binders. Accordingly, it is suggested that studies be conducted with the TSRST of mixes containing such binder before fully implementing the SHRP binder specifications for these materials.

While the asphalt/binder properties (e.g. $G*\sin\delta$) play an important role in the fatigue response of asphalt-aggregate mixes, they alone may not provide sufficiently reliable estimates of fatigue cracking in pavement structures. Accordingly, it is strongly recommended for critical design situations (e.g. heavy traffic volumes and/or the use of modified binders) that mix fatigue testing of the type described herein be conducted.

Aggregate characteristics have a significant effect on the resistance of mixes to permanent deformation. Thus, the correlations between $G*/\sin\delta$ of the binder appearing in the binder specification and the various measures of permanent deformation used herein were generally poor. In cases where mix characteristics are such that interparticle friction is low and the mix is subjected to harsh loading and high temperature conditions, the parameter $G*/\sin\delta$ may be used to screen binders which may provide inferior performance. Generally the results underscore the importance of mix testing, in addition to binder testing, for evaluation of permanent deformation in pavements.

The limited testing performed on modified binder in both fatigue and permanent deformation suggest that the tests developed in the A-003A are suitable to define response of mixes containing these types of binders as well as conventional asphalts.

In general the results of this investigation indicate that asphalt properties as well as aggregate properties will influence mix performance and underscore that mix characteristics be evaluated by testing to insure confidence in their effects on pavement performance.

REFERENCES

[1] Tayebali, A. A., Tsai, B., and Monismith, C. L., *Stiffness of Asphalt-Aggregate Mixes*, Report No. SHRP-A-388, Strategic Highway Research Program, National Research Council, Washington, D.C., 1994.

[2] Tayebali, A. A., Deacon, J. A., Coplantz, J. S., Harvey, J. T., Finn, F. N., and Monismith, C. L., *Fatigue Response of Asphalt-Aggregate Mixes*, Report No. SHRP-A-404, Strategic Highway Research Program, National Research Council, Washington, D.C., 1994.

[3] Sousa, J. B., Deacon, J. A., Weissman, S., Harvey, J. T., Monismith, C. L., Leahy, R. B., Paulsen, G., and Coplantz, J. S., *Permanent Deformation Response of Asphalt-Aggregate Mixes*, Report No. SHRP-A-414, Strategic Highway Research Program, National Research Council, Washington, D.C., 1994.

[4] University of California at Berkeley, Oregon State University, Austin Research Engineers, Inc., and SWK Pavement Engineering, *Stage 1 Validation of the Relationship Between Asphalt Properties and Asphalt-Aggregate Mix Performance*, Report No. SHRP-A-398, Strategic Highway Research Program, National Research Council, Washington, D.C., 1994.

[5] University of California at Berkeley, Oregon State University, and Austin Research Engineers, Inc., *Accelerated Performance Tests for Asphalt-Aggregate Mixes and Their Use in Mix Design and Analysis Systems*, Report No. SHRP-A- , Strategic Highway Research Program, National Research Council, Washington, D.C., 1994.

[6] Jung, D-H. and Vinson, T. S., *Low Temperature Cracking — Test Selection*, Report No. SHRP-A-400, Strategic Highway Research Program, National Research Council, Washington, D.C., 1994.

[7] Jung, D-H. and Vinson, T. S., *Validation of the A-002A Findings for Low-Temperature Cracking*, Report No. SHRP-A-399, Strategic Highway Research Program, National Research Council, Washington, D.C., 1994.

[8] Kanerva, H. K., Vinson, T. S., and Zeng, H., *Low Temperature Cracking — Field Validation for the TSRST*, Report No. SHRP-A-401. Strategic Highway Research Program, National Research Council, Washington, D.C., 1994.

[9] Bell, C. A., AbWahab, Y., Christi, M. E., and Sosnovske. D., *Selection of Laboratory Aging Procedures for Asphalt-Aggregate Mixtures*, Report No. SHRP-A-383, Strategic Highway Research Program, National Research Council, Washington, D.C., 1994.

[10] Bell, C. A. and Sosnovske, D., *Aging: Binder Validation*, Report No. SHRP-A-384, Strategic Highway Research Program, National Research Council, Washington, D.C., 1994.

[11] Bell, C. A., Wieder, A. J., and Fellin, M. J., *Laboratory Aging of Asphalt-Aggregate Mixtures Field Validation*, Report No. SHRP-A-390, Strategic Highway Research Program, National Research Council, Washington, D.C., 1994.

[12] Terrel, R. L. and Al-Swailmi, S., *Water Sensitivity of Asphalt-Aggregate Mixtures Test Development*, Report No. SHRP-A-403, Strategic Highway Research Program, National Research Council, Washington, D.C., 1994.

[13] Scholz, T., Terrel, R. L., Al-Joaib, A., and Bea, J., **Water Sensitivity: Binder Validation**, Report No. SHRP-A-402, Strategic Highway Research Program, National Research Council, Washington, D.C., 1994.

[14] Allen, W. L. and Terrel, R. L., **Field Validation of the Environmental Conditioning System**, Report No. SHRP-A-396, Strategic Highway Research Program, National Research Council, Washington, D.C., 1994.

[15] Anderson, D. A., Christensen, D. W., Bahia, H. V., Dongre, R., Sharma, M. G., Antle, C. E., and Button, J., **Binder Characteristics and Evaluation, Volume 3**, Report No. SHRP-A-369, Strategic Highway Research Program, National Research Council, Washington, D.C., 1994.

[16] Hicks, R. G., Finn, F. N., Monismith, C. L., and Leahy, R. B., "Validation of SHRP Binder Specification Through Mix Testing," *Journal*, Association of Asphalt Paving Technologists, Vol. 62, 1993, pp. 565-614.

Reynaldo Roque,[1] Dennis H. Hiltunen,[2] William G. Buttlar,[3] and Tariq Farwana[4]

DEVELOPMENT OF THE SHRP SUPERPAVE MIXTURE SPECIFICATION TEST METHOD TO CONTROL THERMAL CRACKING PERFORMANCE OF PAVEMENTS

REFERENCE: Roque, R., Hiltunen, D. H., Buttlar, W. G., and Farwana, T., "**Development of the SHRP Superpave Mixture Specification Test Method to Control Thermal Cracking Performance of Pavements,**" Engineering Properties of Asphalt Mixtures and the Relationship to their Performance, ASTM STP 1265, Gerald A. Huber and Dale S. Decker, Eds., American Society for Testing and Materials, Philadelphia, 1995.

ABSTRACT: The indirect tensile creep and failure test at low temperatures (ITLT) was selected by SHRP to control thermal cracking performance within the SUPERPAVE mixture design and analysis system. Fundamental viscoelastic properties and fracture parameters obtained from the test are used in the SUPERPAVE thermal cracking model to predict thermal cracking performance (cracking as a function of time) of asphalt pavements of variable thicknesses in different temperature regimes. This approach gives pavement and mixture designers the capability of determining not only that one mixture is better than another, but also quantifies how much better one mixture is than another in terms of its cracking performance. The background and principles used to identify this test method are presented in this paper along with the results of analytical and laboratory work conducted to identify specific test procedures. Comparisons between cracking predicted using ITLT test results and observed cracking in over 35 test sections in the United States and Canada have indicated that this test is suitable for control of thermal cracking of asphalt mixtures.

KEYWORDS: Indirect tension, thermal cracking, low temperature cracking, asphalt mixture, creep compliance, relaxation modulus, stiffness, master curve.

---

[1]Associate Professor, Department of Civil Engineering, University of Florida, Gainesville, FL 32611.
[2]Associate Professor, Department of Civil and Environmental Engineering, Penn State University, University Park, PA 16802.
[3]Dwight D. Eisenhower Graduate Research Fellow and Ph.D. Candidate, Penn State University, University Park, PA 16802.
[4]Consultant, Foreign Trade, Ltd., Dublin, Ohio.

INTRODUCTION

Two of the primary objectives of the asphalt program within the Strategic Highway Research Program (SHRP) were:

- To develop performance-based specifications for asphalt binders and mixtures.

- To develop performance prediction models to support the specifications.

Successful development of performance-based specifications and models would allow designers to produce mixtures that meet specific levels of performance for a given pavement structure and environmental conditions. Designers would also be able to compare mixtures based on their performance as opposed to simply comparing properties whose relationships to performance are unclear. The implication is that a quantitative comparison of performance would allow designers to determine not only that one mixture is better than another, but also how much better one mixture is than another mixture.

In terms of thermal cracking, performance is defined as the amount or frequency of transverse cracking as a function of time or age of the pavement. Therefore, in order to meet the objectives of the SHRP asphalt program, a test and analysis system was needed that accounted for mixture properties and mechanisms that control thermal cracking. At the same time, the test had to be simple enough for purposes of mixture design.

This paper describes the efforts of the authors to identify and develop a mixture testing system that meets the requirements stipulated above. Work conducted to validate the suitability of the testing system selected is also summarized.

SELECTION OF TESTING SYSTEM

System Requirements

A mechanics-based approach, as opposed to an empirical approach, was selected by the authors to address the needs and goals of the SHRP asphalt program. The idea was to establish a system that would use the principles of mechanics to directly account for the relevant factors that influence thermal cracking. For example, thermal stresses within the asphalt concrete surface layer are a function of asphalt concrete stiffness and coefficient of thermal contraction, pavement temperature, cooling rate, thermal loading history, and depth within the asphalt concrete layer. It would be difficult, if at all possible, to account for the effects of each of these factors on thermal stresses and thermal cracking performance by using an empirical approach. A very large experiment would be required to obtain a broad enough data base to reliably determine empirical coefficients that account for the effects of these variables and their interactions. However, since the mechanism of thermal stress development in viscoelastic materials is well understood, one can directly account for these factors if the fundamental viscoelastic properties of the asphalt mixture are known. Thus, by using mechanics, the effects of all factors mentioned are accounted for in the thermal stress prediction, while these factors would have to be either ignored or treated separately if an empirical approach were used.

The selection of a mechanics-based approach required the use of a testing system that provided the fundamental asphalt mixture properties that control thermal cracking of asphalt pavements. The mixture properties required were:

- The viscoelastic properties (time- and temperature-dependent relaxation modulus) and coefficient of thermal contraction, which control thermal stress development.

- The fracture parameters, which control the rate of crack development for specified stress conditions.

Given the nature of the problem, tensile properties were required. Therefore, only tests which induced a tensile stress state were considered.

Candidate Tests

The following tests have been used to induce tension in asphalt mixtures:

- Thermal stress restrained specimen test (TSRST).

- Direct tension test.

- Indirect Tension Test

The TSRST *(1)* induces tension by lowering the temperature of a restrained asphalt beam or cylinder. A cooling rate of 10° C/hr is generally used so the test can be completed within a reasonable period of time. In general, however, this cooling rate is far in excess of pavement cooling rates in the field. Stress is measured and plotted as a function of temperature up to the instant of failure. Two mixture parameters are generally obtained from the test: fracture stress and fracture temperature, which correspond to the stress and temperature when failure occurs. The idea is that the lower the fracture temperature, the better the mixture's resistance to low temperature cracking. Other parameters have also been obtained from the test to evaluate the mixture's susceptibility to thermal cracking. The advantages of this test are its intuitive appeal (it appears to model reality) and the fact that coefficient of thermal contraction does not have to be measured or estimated to obtain results.

Unfortunately, fundamental viscoelastic and fracture properties needed to predict performance under a variety of environmental conditions, pavement thicknesses, etc., cannot be determined from this test. Therefore, the authors considered the test to be unsuitable to accomplish the stated objectives.

Tests to Determine Fundamental Tensile Properties

A tensile creep test was considered the most effective way of determining viscoelastic properties of mixtures. Direct tension tests *(2, 3, 4)* involve the use of asphalt concrete beams or cylinders which must be epoxied or otherwise restrained on each end. Unlike steel, which is highly ductile, asphalt concrete is very brittle at low temperatures, so that stress concentrations induced by loading grips or epoxy may result in localized failure. Also, use of epoxy requires setting time (usually long because low modulus epoxy is required for low temperature work). Finally, beams or long cylinders are generally considered more difficult to produce for routine laboratory testing. The primary advantage of direct tension tests is that a uniaxial stress is applied (assuming perfect alignment between loading heads), which makes interpretation very simple.

Indirect tension tests *(5, 6)* involve the use of short cylinders (pills), which are routinely produced in bituminous laboratories or obtained from field cores. Epoxy is not required because a compressive load is applied along a vertical axis to induce tensile stresses along a horizontal axis. The primary disadvantages of indirect tension tests are that a multiaxial (three-dimensional) stress state is applied that makes interpretation difficult, and that high compressive stresses and restraint in the vicinity of the loading platens may result in localized damage which makes platen to platen measurements uninterpretable *(7, 8, 9)*.

It was felt, and later verified, that interpretation and measurement problems associated with conventional indirect tensile testing could be addressed through the development of an appropriate measurement and analysis system. Therefore, the authors felt that the advantages of the indirect tensile test outweighed the disadvantages of direct tension testing. So the indirect tensile testing mode was selected for further development.

Development of Modified Indirect Tensile Test

A new measurement and analysis system was developed to obtain properties accurately using the indirect tensile test. Figure 1 shows a schematic of the measurement system, which features the use of subminiature LVDTs mounted on the center of the specimen's flat face. A similar approach has been used by Anderson *(5)* in Canada to determine low temperature mixture propeties, except that the system used in the SHRP test obtains a vertical as well as a horizontal measurement. It was determined analytically *(7)* that stresses within the central quarter (one quarter of the diameter) of the specimen are unaffected by the effects of the loading platens. Therefore, measurements obtained in this region can be interpreted with confidence.

An analysis procedure was developed to interpret measurements obtained with the new measurements system *(10)*. The procedure accounts for three-dimensional stress effects and corrects for measurement errors induced by specimen

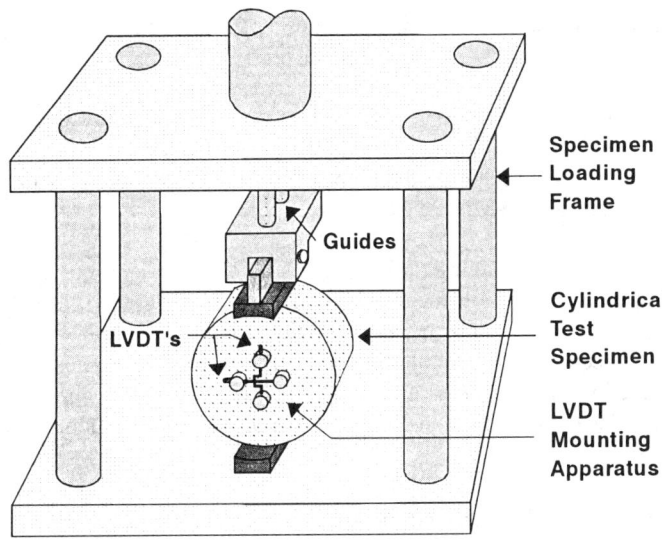

Figure 1. Equipment Schematic

bulging during loading. The system was analytically shown to provide superior accuracy for determining moduli and Poisson's ratio as compared to conventional indirect tensile testing systems that use externally mounted deformation sensors *(7)*. Laboratory investigations showed that measured moduli and Poisson's ratios were not only reasonable, but also followed expected trends when controlled changes were made to the asphalt mixtures tested. As expected, lower stiffnesses and lower Poisson's ratios were determined for mixtures with higher air voids than for mixtures with lower air voids. Also as expected, mixtures produced with stiffer binders were determined to have higher stiffnesses than mixtures produced using lower stiffness binders. In addition, moduli determined using the new indirect tension testing and analysis system agreed with moduli determined from direct tension tests on the same mixtures *(10, 11)*.

DEVELOPMENT OF PRELIMINARY TESTING PROCEDURES

The fundamental properties required to predict thermal cracking are:

- Relaxation modulus, which is time- and temperature-dependent and controls thermal stress development.

- Fracture parameters, which are temperature dependent and control the rate of crack propagation.

Direct measurement of relaxation modulus requires stress relaxation tests, which are more difficult to perform than creep tests. Creep compliance can be measured in lieu of relaxation modulus then transformed mathematically to determine relaxation modulus *(11)*. This was the approach used in the SHRP test procedure.

Since for the thermal loading problem, the load is applied over relatively long periods time, relatively long loading time creep tests would be required to determine compliances that are relevant to this problem, particularly at lower temperatures. Given that the SHRP test needed to be suitable for routine mixture evaluation and design, long loading times were impractical. Therefore, the principles of time-temperature superposition were used to describe the time- and temperature-dependent mixture compliance.

Principles of Time-Temperature Superposition

The principles of time-temperature superposition are illustrated in figure 2. For linear viscoelastic materials there is a correspondence between loading time and temperature that is defined by a shift factor (a_T)-temperature relationship. This relationship can be established experimentally by conducting creep tests at multiple temperatures then shifting the resulting compliance curves to obtain one smooth continuous curve at a single temperature (refered to as the reference temperature). The resulting curve is called the master creep compliance curve. The following time-dependent creep compliance function was used to define the master creep compliance curve:

$$D(\xi) = D(0) + \sum_{i=1}^{N} D_i(1 - e^{-\xi/\tau_i}) + \frac{\xi}{\eta_v}$$

where: $D(\xi)$ = creep compliance at reduced time ξ
ξ = reduced time = t/a_T
a_T = temperature shift factor
$D(0), D_i, \tau_i, \eta_v$ = Prony series parameters

Prony series parameters and temperature shift factors are determined by fitting the data using a nonlinear regression routine *(11)*.

Successful and accurate development of the master compliance curve and corresponding shift factor-temperature relationship requires that there be overlap between compliances obtained at successive test temperatures. In other words, the compliance at

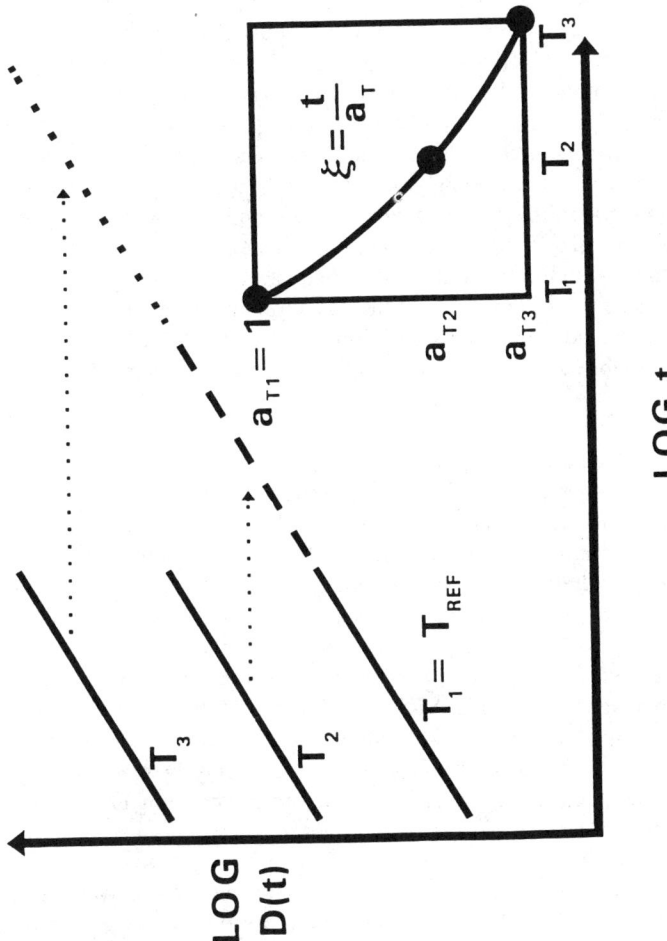

Figure 2: Time-Temperature Superposition

the longest loading time for test temperature T_1 in figure 2 should be greater than or equal to (or at least close to) the compliance at the shortest loading time for test temperature T_2.

Analyses to Select Loading Times and Temperatures

Analytical studies were conducted to identify the most effective combination of test temperatures and loading times to determine master compliance curves of mixtures at low temperatures. Potential combinations are infinite, but several guidelines were used to limit the possibilities. One guideline was to minimize the number of test temperatures, since changing test temperatures is the most time consuming portion of testing. In general, it was determined that a minimum of three test temperatures were needed, since this is the minimum number required to define a nonlinear shift factor-temperature relationship. Therefore, the goal of the analytical study was to identify the most effective temperature increment between the three test temperatures, and the loading time required at each test temperature to get sufficient overlap to reliably define the master compliance curve. Test temperatures were limited to less than or equal 0° C to address low temperature thermal cracking and to maximize the chances that mixture response during testing would remain in the linear range. Also, a minimum temperature increment of 10° C was considered reasonable, since this would provide mixture data over a 20° C temperature range.

Figures 3 and 4 show creep compliance curves at 10° and 15° C temperature increments, respectively, generated from binder-to-mixture stiffness relationships established by Heukelom and Klomp *(12)*. Binder stiffnesses from the AAG-1 asphalt cement, which is one of the stiffest binders in the SHRP materials reference library (MRL), were used along with typical mixture volumetrics to generate the curves. As shown in figure 3, less than 1000 seconds (16.7 minutes) of creep data were required to get sufficient overlap to generate the master curve using 10° C increments, while nearly 10,000 seconds (167 minutes, 2.75 hours) were required when a 15° C increment was used. Since testing experience had indicated that a minimum of three replicate specimens were required to obtain properties reliably with the new indirect tensile testing system, nine specimens were required for mixture evaluation and design if three test temperatures were used. At 2.75 hours per specimen, nearly 26 hours of creep testing alone would be required if a temperature increment of 15° C were used.

Based on these analyses, test temperatures of 0°, -10°, and -20° C were selected along with a loading time of 1000 seconds at each test temperature for use in determining the master compliance curve of asphalt mixtures. Similar analyses were also conducted with other SHRP MRL asphalts. It was determined that the selected temperatures and loading times were suitable for the other asphalts, since these were generally less stiff and less temperature susceptible than the AAG-1 asphalt.

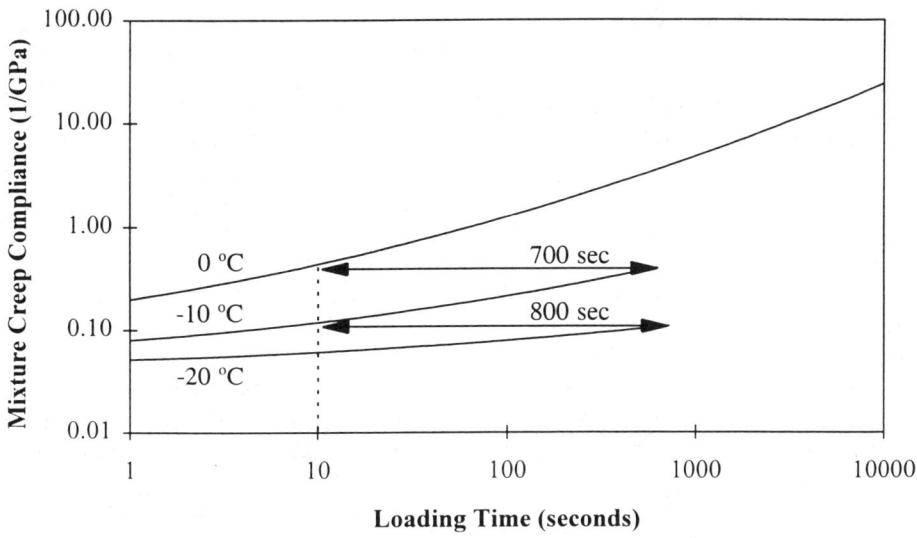

Figure 3. Generated Mixture Creep Compliance for 10 °C Temperature Increments: SHRP AAG-1 Asphalt.

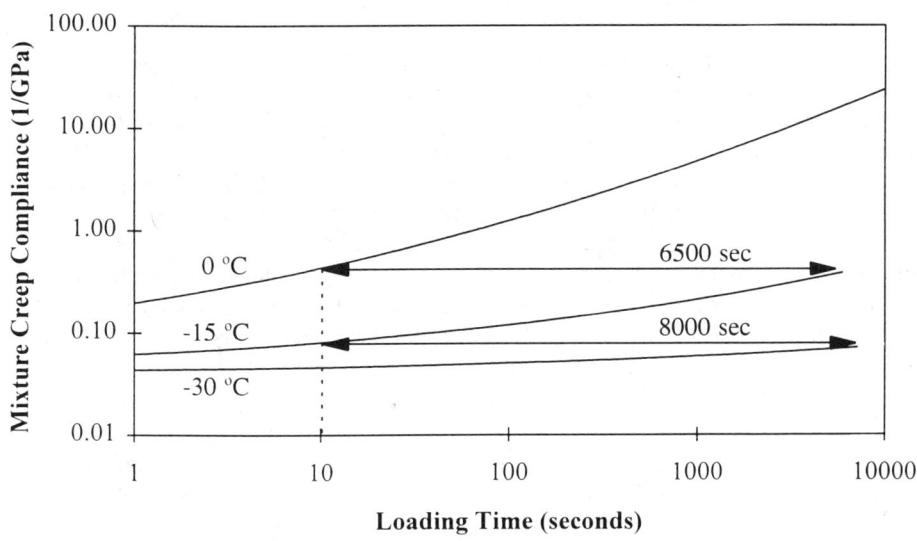

Figure 4. Generated Mixture Creep Compliance for 15 °C Temperature Increments: SHRP AAG-1 Asphalt.

Preliminary Laboratory Tests

Laboratory tests were conducted on a dense-graded surface course mixture to identify strain levels within which the mixture response could be considered to be within the linear viscoelastic range. The data were also used to verify the suitability of a 10° C temperature increment and 1000-second loading time to generate the master mixture compliance curve.

Allowable strain levels--Figure 5 shows mixture compliance obtained at three stress levels (20, 25, and 30 psi tensile stress) and two temperatures (-5° and -15° C). The figure shows that for loading times of up to 10,000 seconds, the compliances determined from the three stress levels were essentially identical for all three stress levels used. The implication is that mixture response was linear viscoelastic up to the maximum strain level of approximately 2000 microstrain (mm/mm x 10^{-6}) applied during these tests.

Since the linear viscoelastic range of other mixtures may be lower than 2000 microstrain, a lower strain level of 500 microstrain was recommended as the maximum allowable for the SHRP indirect tensile creep test. However, for most mixtures results should be well within the linear viscoelastic range even if this recommended minimum value is exceeded. Since the measurement system to be used for the SHRP indirect tensile creep and failure test calls for precision of ±10 microstrain (actually, ±250 x 10^{-6} mm over a 25 mm gage length) a minimum strain level of 200 microstrain was recommended to obtain accurate results.

Verification of loading times and temperatures--Figure 6 shows measured mixture compliances obtained using a temperature increment of 10° C and a loading time of 1000 seconds. The figure clearly shows that these data provided sufficient overlap to accurately shift the compliance data to determine the master compliance curve. Based on these preliminary findings, the decision was made to use 1000 second creep tests at a temperature increment of 10° C to evaluate the effectiveness of the indirect tensile creep and failure test in the field. The procedure was also used to evaluate the test method by comparing fracture temperatures predicted from the test results to those determined from the thermal stress restrained specimen test (TSRST). Details of these comparisons may be found in the report by Lytton et al. *(11)*. The correspondence between predicted and measured fracture temperatures was excellent.

Fracture Parameters

The amount of crack propagation induced by a given thermal cooling cycle can be predicted by using the Paris law of crack propagation:

$$\Delta C = A(\Delta K)^n$$

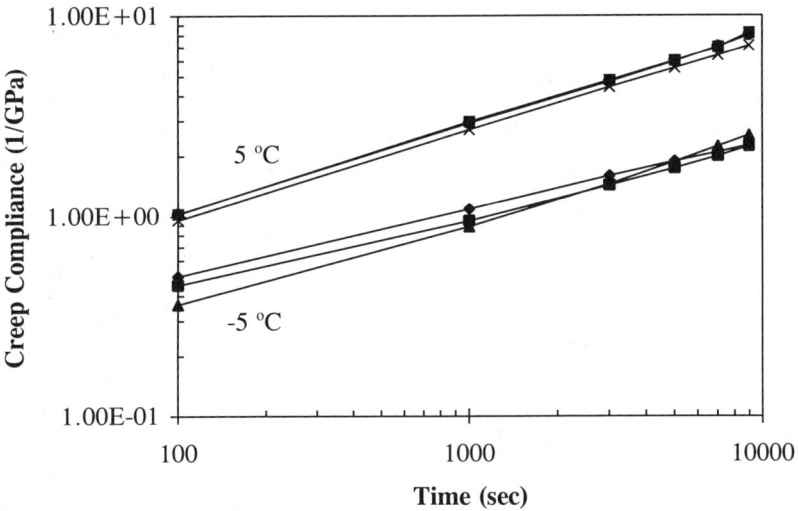

Figure 5. Creep Compliances from Indirect Tensile Tests Performed at 3 Stress Levels and 2 Tempertures.

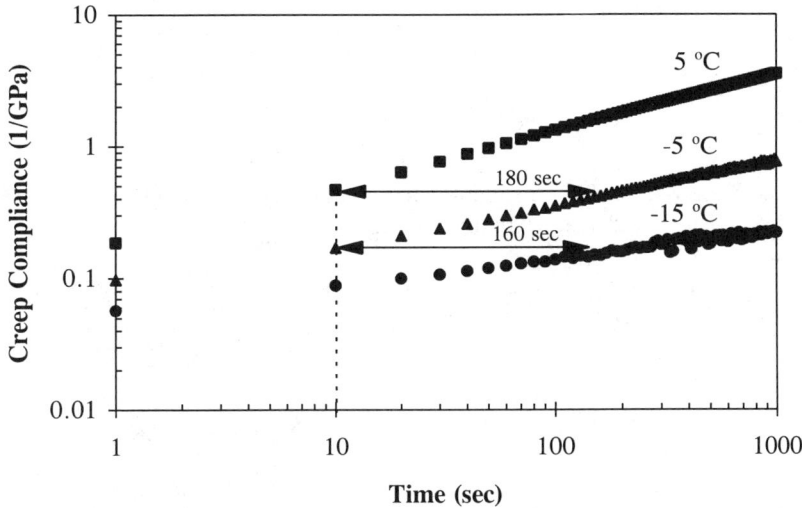

Figure 6. Creep Compliances for Pennsylvania Surface Mixture (ID-2) at 3 Temperatures.

where:
- ΔC = change in the crack depth due to a cooling cycle
- ΔK = change in the stress intensity factor due to a cooling cycle
- A, n = fracture parameters for the asphalt mixture

Schapery's theory *(13, 14)* of crack propagation in viscoelastic materials indicates that the fracture parameters A and n are theoretically related to the following fundamental material properties:

- The slope of the linear portion of the log compliance-log time master curve determined from creep tests.

- The undamaged tensile strength of the material.

- The fracture energy density of the material determined experimentally by monitoring the energy release through crack propagation.

The m-value and the tensile strength of the material are obtained directly from the indirect tensile creep and failure test *(11)*. However, determination of the fracture energy density would require additional and fairly complex testing, that could not be incorporated into a mixture specification scenario. Fortunately, experimental results *(15, 16)* have shown that reasonable estimates of A and n can be obtained from the m-value and the tensile strength of the material, which agrees with the theoretical work of Schapery. Based on this work, the following relationship, which is a modified version of the relationship developed experimentally by Molenaar *(15)*, was established to determine the parameter A:

$$\log A = 4.389 - 2.52 * \log(K * \sigma_m * n)$$

where: K = coefficient determined through field calibration = 10,000
σ_m = undamaged mixture strength

The following equation, which was developed experimentally by Lytton et al. *(16)*, was used to determine the parameter n:

$$n = 0.8 * (1 + \frac{1}{m})$$

FIELD EVALUATION (SHRP GPS TEST SECTIONS)

The primary focus of the SHRP A005 research contract was to validate the binder and mixture specification tests using field data. For thermal cracking, the objective was to determine whether the indirect tensile creep and failure test could be used to effectively control thermal cracking in the field. This was accomplished by comparing thermal cracking predicted using laboratory test results to observed thermal cracking in the field. Field cores were obtained from 23 test sections exhibiting variable performance in different environmental regions within the United States. Indirect tensile creep and failure tests were conducted on triplicate specimens at three test temperatures (0°, -10°, and -20° C). A loading time of 1000 seconds was used. The results of the validation, which have been published elsewhere *(10, 16)*, indicated that the indirect tensile creep and failure test is clearly suitable to support the SHRP mixture specification for thermal cracking. Excellent correspondence was obtained between observed thermal cracking in the field and thermal cracking predicted using fundamental properties determined from the test.

The extensive amount of testing performed on the broad range of mixtures involved in the field investigation also allowed for further evaluation of the testing procedure. In general, the testing procedure was found to be suitable for the majority of mixtures evaluated. In other words, the creep data obtained provided sufficient overlap to accurately shift the compliance curves to determine the master compliance curves of most mixtures tested. However, in a few cases, the 10° C increments and 1000-second loading times did not provide the overlap necessary for accurate shifting. One such case is illustrated in figure 7. As shown in the figure, an additional creep test at -15° C was needed to obtain the data necessary for accurate shifting.

For research purposes, this was not a major problem. However, this is a serious problem from the standpoint of having a fully automated procedure to generate mixture data for performance evaluation, as required by the SHRP SUPERPAVE system. If there is insufficient overlap between compliance curves determined at different temperatures, then the nonlinear regression routines used to generate the master compliance curve will very likely become unstable and will not converge. A technician working with SUPERPAVE would not know why this happened or how to resolve the problem. Therefore, there was a need to develop a modified procedure to generate compliance data that would guarantee sufficient overlap such that the master compliance curve could be determined reliably. At the same time, reduced testing time would also be desirable.

DEVELOPMENT OF MODIFIED TESTING PROCEDURES

Two things prompted development of a modified testing and analysis procedure: 1) the need to guarantee sufficient overlap between compliance data obtained at different temperatures; and 2) the need to minimize testing time to make the SHRP test as practical as possible for routine mixture evaluation and design. Both things could be accomplished by making use of the binder stiffness data obtained from the bending beam rheometer test

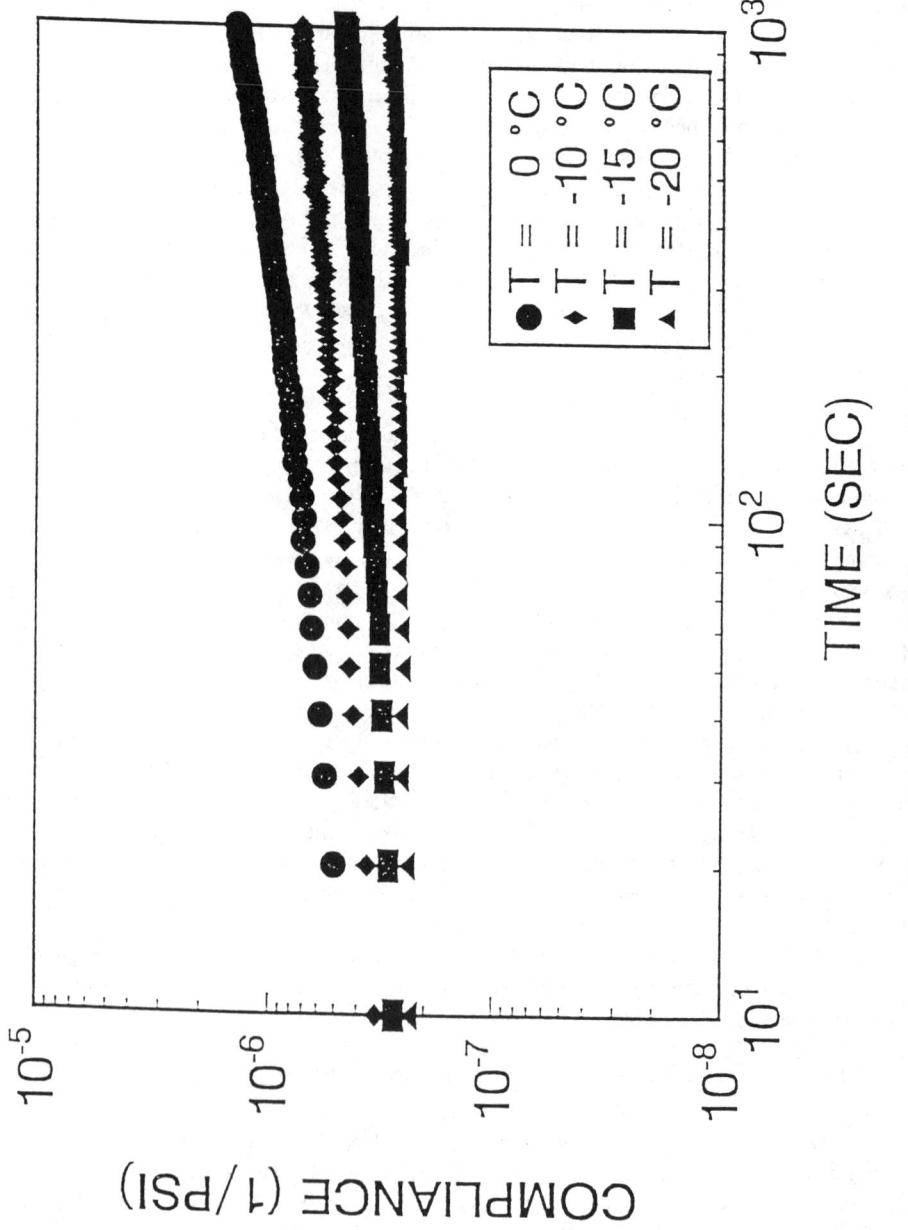

Figure 7: Mixture compliance curves at multiple temperatures.

(18). Prior work by numerous researchers *(e.g. 6, 12)* has indicated that binder and mixture stiffness are strongly related. This strong correspondence was also observed for the binders and mixtures tested as part of the SHRP field validation process described above.

Figure 8 shows a plot of log mixture stiffness (determined as the inverse of the measured creep compliance) as a function of log binder stiffness as determined from bending beam rheometer tests performed on binder extracted from mixture obtained from one of the field test sections. Clearly, the two are strongly related and one could easily establish a linear relationship between the two by use of simple linear regression. The implication of having this relationship is that one is able to generate mixture compliances (inverse of stiffness) at any temperature and loading time once the binder stiffnesses are defined using the bending beam rheometer test.

It was also observed that essentially the same linear relationship could be established even if most of the mixture data were eliminated from the plot in figure 8. Three groups of data can be identified in figure 8, corresponding to each of the three test temperatures at which mixture tests were performed (0°, -10°, and -20° C). The data within each group corresponds to stiffnesses (inverse of compliance) at different loading

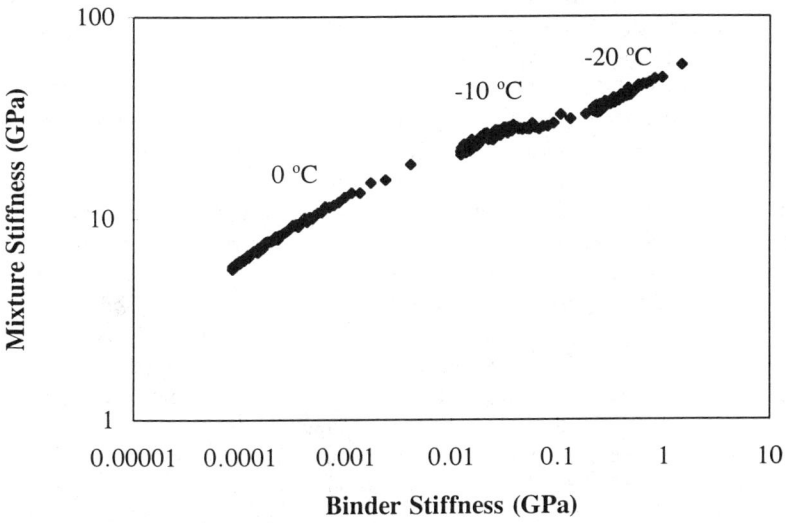

Figure 8. Log Mixture Stiffness Versus Log Binder Stiffness.

times, up to the maximum loading time of 1000 seconds. So the minimum stiffness on the plot corresponds to a temperature of 0° C and a loading time of 1000 seconds, while the maximum stiffness on the plot corresponds to a temperature of -20° C and a loading time of 1 second. If one eliminated (or did not obtain) the last 900 seconds of data from the mixture compliance test at each of the three test temperatures, one would still be able to generate the relationship between binder and mixture stiffness.

Based on these observations, it was recommended that loading times for the creep test portion of the SHRP indirect tensile creep and failure test be reduced from 1000 to 100 seconds. A computational scheme was developed and included in SUPERPAVE that determines the binder-to-mixture stiffness relationship for each mixture from which a series of mixture compliance curves are generated for development of the master compliance curve. This procedure guarantees that the overlap between compliance curves at different temperatures is sufficient for accurate and reliable generation of the master compliance curve.

FIELD VALIDATION OF MODIFIED TESTING PROCEDURE

An independent set of field data were used to validate the modified testing procedure (i.e. 100-second creep tests combined with binder data to generate the mixture compliance curve). Fourteen field test sections were constructed in 1991 and 1992 as part of the Canadian Strategic Highway Research Program (C-SHRP) to examine low temperature transverse cracking of asphaltic concrete pavements in Canada *(19)*. This research project, entitled "Performance Correlation for Quality Paving Asphalts", was initiated in 1989. One objective of the study was to correlate low temperature cracking to the new Canadian General Standards Board specification. The study was also used to evaluate the SHRP binder and mixture tests for control of thermal cracking in Canada. Test sections were controlled and carefully monitored in Alberta (7 sections), Ontario (3 sections), and Quebec (4 sections) using a range of binders typically used in Canada, as well as some that were specifically refinery manufactured as premium asphalt cements/. Field cores obtained from the test sections before the third winter were tested to determine fundamental mixture properties (master mixture compliance curves and fracture parameters) using the SHRP indirect tensile creep and failure test. These properties were used along with site-specific temperature and pavement structure data to predict the thermal cracking performance (amount of cracking vs. time) of each test section using the thermal cracking model developed for SHRP and included in SUPERPAVE. Thermal cracking predictions were compared to observed thermal cracking after the third winter to evaluate the adequacy of the SHRP SUPERPAVE system. Details on the predictions and evaluations can be found in the CTAA publication by Roque and Hiltunen *(20)* and in future reports to be published by C-SHRP. The relevant findings from the investigation are:

- Eleven of the 14 thermal cracking predictions (79%) made using 100-second creep test data were in good agreement with observed cracking after three winters of service. It is important to note that these comparisons were based on blind predictions using a model that was not calibrated with Canadian data.

- In general, thermal cracking predictions made using 1000-second creep test data were in agreement with predictions made using 100-second creep test data. However, in two of the fourteen test sections, the procedures resulted in significantly different predictions.

Additional work is needed to determine the source of the discrepancy between the two data interpretation procedures and to develop an interpretation procedure that provides consistent results in all cases, if possible. This work is currently being pursued by the authors, the focus being to improve the relationship between binder and mixture stiffness, which is suspected to be the primary source of the observed differences.

SUMMARY AND CONCLUSIONS

A practical testing and analysis procedure was developed to determine the fundamental viscoelastic and fracture properties required to predict the thermal cracking performance of asphalt mixtures. The procedure (the indirect tensile creep and failure test at low temperatures) is now part of the SHRP SUPERPAVE mixture design and analysis system. The test and associated SUPERPAVE system give pavement and mixture designers the capability of determining not only that one mixture is better than another, but also quantifies how much better one mixture is than another in terms of its thermal cracking performance.

Thermal cracking predictions based on mixture properties determined from the test and the SUPERPAVE thermal cracking model have been found to be in excellent agreement with observed thermal cracking in over 35 field test sections in the United States and Canada. Also, material properties determined with the test have exhibited expected trends when mixture characteristics such as asphalt cement type and air void content were varied. Therefore, it appears that SHRP indirect tensile creep and failure test is suitable for control of thermal cracking of asphalt mixtures.

Some discrepancies were found between thermal cracking predictions made using 1000-second creep test data and predictions from 100-second creep test data. The authors are currently pursuing the source of these discrepancies and the development of modified data procedures that will eliminate these discrepancies.

REFERENCES

(1) Jung, D. and T.S. Vinson, 1993. "Low Temperature Cracking Resistance of Asphalt Concrete Mixtures". *Journal of the Association of Asphalt Paving Technologists,* Vol. 62, pp. 54-92.

(2) Fromm, H.J. and W.A. Phang, 1972. "A Study of Transverse Cracking of Bituminous Pavements." *Proceedings of the Association of Asphalt Paving Technologists*, Vol. 41, pp. 383-423.

(3) Haas, R., 1973. "A Method for Designing Asphalt Pavements to Minimize Low-Temperature Shrinkage Cracking," RR-73-1, The Asphalt Institute.

(4) Kallas, B.F., 1982. "Low-Temperature Mechanical Properties of Asphalt Concrete." RR-82-2, The Asphalt Institute.

(5) Anderson, K.O., S.C. Leung, S.C. Poon, and K. Hadipour, 1986. "Development of a Method to Evaluate the Low Temperature Tensile Properties of Asphalt Concrete", *Proceedings of the Canadian Technical Asphalt Association*, Vol. 31, pp. 156-188.

(6) Ruth, B. E. and J. D. Maxfield, "Fatigue of Asphalt Concrete." Final Report-Project 245-D54, Department of Civil Engineering, University of Florida, November 1977.

(7) Roque, R. and W.G. Buttlar, 1992. "The Development of a Measurement and Analysis System to Accurately Determine Asphalt Concrete Properties Using the Indirect Tensile Mode." *Journal of the Association of Asphalt Paving Technologists*, Vol. 61, pp. 304-332.

(8) Mamlouk, M. S. and Sarofim, "The Modulus of Asphalt Mixtures--An Unresolved Dilemma." Presented at the 67th Annual Meeting of the Transportation Research Board, 1988.

(9) Sousa, J. B., R. Taylor, and A. J. Tanco, "Analysis of Some Laboratory Testing Systems for Asphalt-Aggregate Mixtures" Presented at the 70th Annual Meeting of the Transportation Research Board, 1991.

(10) Buttlar, W. B. and R. Roque, 1994. "Experimental Development and Evaluation of the New SHRP measurement and Analysis System for Indirect Tensile Testing of Asphalt Mixtures at Low Temperatures," Presented at the annual meeting of the Transportation Research Board, and recommended for publication in future *Transportation Research Record* (1994).

(11) Lytton, R. L., R. Roque, J. Uzan, D. R. Hiltunen, E. Fernando and S. M. Stoffels, 1993. "Performance Models and Validation of Test Results," *Final Report to Strategic Highway Research Program; Asphalt Project A-005*, 529 pp.

(12) Heukelom, W., and A.J.G. Klomp, 1964. "Road Design and Dynamic Loading." *Proceedings of the Association of Asphalt Paving Technologists*, Vol. 33, pp. 92-125.

(13) Schapery, R.A., 1984. "Correspondence Principles and a Generalized J Integral for Large Deformation and Fracture Analysis of Viscoelastic Materials." *International Journal of Fracture*, Vol. 25, pp. 195-223.

(14) Schapery, R.A., 1986. "Time-Dependent Fracture: Continuation Aspaects of Crack Growth." *Encyclopedia of Materials Science and Engineering*, M.B. Bever (Ed.), Pergamon Press, Elmsford, NY, pp. 5043-5053.

(15) Molenaar, A.A.A., 1983. "Structural Performance and Design of Flexible Road Constructions and Asphalt Concrete Overlays." *Ph.D. Dissertation*, Delft University of Technology, Netherlands.

(16) Lytton, R.L., U. Shanmugham, and B.D. Garrett, 1983. "Design of Asphalt Pavements for Thermal Fatigue Cracking." Research Report 284-4, Texas Transportation Institute.

(17) Roque, R., D.R. Hiltunen, and S.M. Stoffels, 1993. "Field Validation of SHRP Asphalt Binder and Mixture Specification Tests to Control Thermal Cracking Through Performance Modeling." *Journal of the Association of Asphalt Paving Technologists*, Vol. 62, pp. 615-638.

(18) Bahia, H.U. and D.A. Anderson, 1992. "The Bending Beam Rheometer; A Simple Device for Measuring Low-Temperature Rheology of Asphalt Binders." *Journal of the Association of Asphalt Paving Technologists*, Vol. 61, pp. 117-153.

(19) *Draft Construction Report: Test Roads, C-SHRP Project, Performance Correlation of Quality Asphalt Paving.* 1993. EBA Engineering, Ltd., Edmonton, Alberta.

(20) Roque, R. and D.R. Hiltunen, 1994. "Use of Canadian SHRP Test Sections to Evaluate the SHRP Indirect Tensile Creep and Failure Test for Control of Thermal Cracking." *Proceedings of the Canadian Technical Asphalt Association*, Vol. XXXIX, pp. 441-464.

Dennis R. Hiltunen[1] and Reynaldo Roque[2]

THE USE OF TIME-TEMPERATURE SUPERPOSITION TO FUNDAMENTALLY CHARACTERIZE ASPHALTIC CONCRETE MIXTURES AT LOW TEMPERATURES

REFERENCE: Hiltunen, D. R. and Roque, R., **"The Use of Time-Temperature Superposition of Fundamentally Characterize Asphaltic Concrete Mixtures at Low Temperatures,"** Engineering Properties of Asphalt Mixtures and the Relationship to their Performance, ASTM STP 1265, Gerald A. Huber and Dale S. Decker, Eds., American Society for Testing and Materials, Philadelphia, 1995.

ABSTRACT: An analytical approach for the determination of the viscoelastic properties of asphaltic concrete mixtures at low temperatures is described. The properties are derived from laboratory test data obtained from the mixture test selected by the Strategic Highway Research Program (SHRP) to support the new mixture specification for thermal cracking. It is shown that the principle of time-temperature superposition can adequately represent the time- and temperature-dependent stiffness properties of asphaltic concrete mixtures at low temperatures. In addition, it is shown that a generalized Maxwell model (Prony series) can accurately represent the master creep compliance curve of asphaltic concrete mixtures at low temperatures.

KEYWORDS: asphaltic concrete, creep compliance, low temperatures, Maxwell model, relaxation modulus, stiffness, thermal cracking, viscoelastic

[1]Associate Professor, Department of Civil Engineering, The Pennsylvania State University, University Park, Pennsylvania 16802.

[2]Associate Professor, Department of Civil Engineering, University of Florida, Gainesville, Florida 32611.

INTRODUCTION

The work presented in this paper was conducted at the Pennsylvania Transportation Institute (PTI) of the Pennsylvania State University (Penn State) as part of the Strategic Highway Research Program (SHRP) A-005 asphalt research contract. The Texas Transportation Institute (TTI) of Texas A&M University was the prime contractor on this project.

The two primary goals of the SHRP A-005 research effort at Penn State were:

- To validate the effectiveness of the new SHRP binder and mixture specification tests to control thermal cracking performance of asphaltic concrete pavements in the field.

- To develop a pavement performance prediction model to support the new SHRP mixture specifications and for use with the SHRP SUPERPAVE software.

In addition, the mixture specification test selected by SHRP to support the mixture specification for thermal cracking was developed at Penn State as part of this contract.

The results of the SHRP A-005 field validation effort related to thermal cracking performance have been presented by Roque, Hiltunen, and Stoffels [1]. Conclusions related to the effectiveness of the new SHRP binder and mixture specifications to control thermal cracking were presented based on the results of the validation effort. A detailed description of the new performance prediction model for thermal cracking has been presented by Hiltunen and Roque [2], and the new mixture testing system for thermal cracking has been described by Roque and Buttlar [3] and Buttlar and Roque [4].

The new mechanics-based thermal cracking performance model predicts the amount (or frequency) of thermal cracking as a function of time. Two of the primary inputs to the model characterize the viscoelastic properties of the asphaltic concrete mixture: 1) the master relaxation modulus relationship, and 2) the slope of the linear portion of the master creep compliance curve on a logarithmic scale. Each of these inputs is derived from the new SHRP mixture test. The purpose of this paper is to describe the analytical approach developed for the determination of these inputs from laboratory test data obtained from the new mixture test.

THERMAL CRACKING MODEL

As described by Haas et al. [5], Roque, Hiltunen, and Stoffels [1], Hiltunen and Roque [2], and others, the primary mechanism leading to thermal cracking is shown in figure 1. Contraction strains induced by cooling lead to thermal tensile

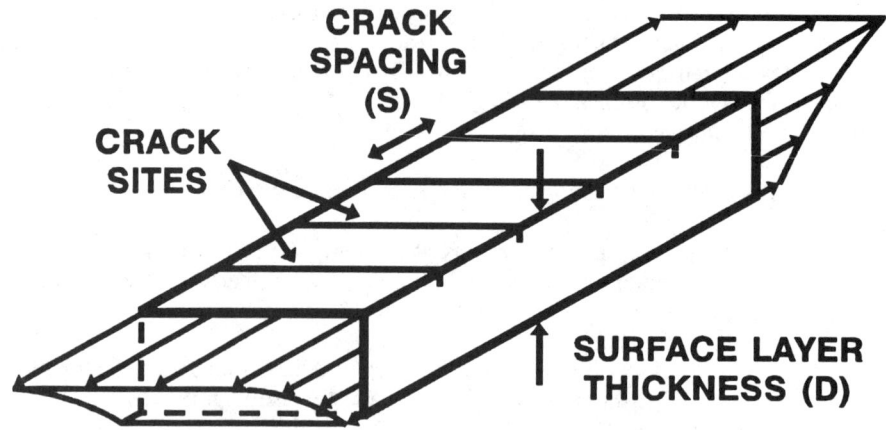

Figure 1. Schematic of Physical Model of Pavement Section

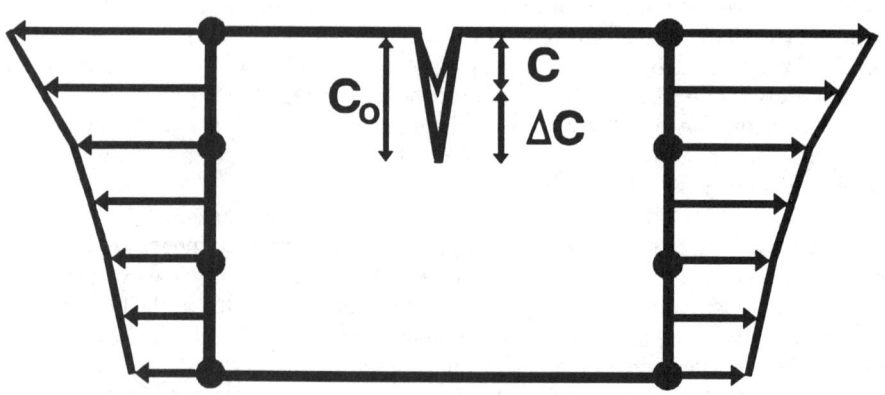

Figure 2. Schematic of Crack Depth (Fracture) Model

stress development in the restrained surface layer. Thermal stress development is greatest in the longitudinal direction of the pavement because there is greater restraint length in that direction. Also, thermal stresses are greatest at the surface of the pavement because pavement temperature is lowest at the surface. Depending upon the magnitude of these stresses and the asphalt mixture's resistance to fracture (crack propagation), transverse cracks may develop at different points along the length of the pavement.

Physical Model

The physical representation of the actual pavement structure assumed in the thermal cracking model is shown in figures 1 and 2. In figure 1 an asphaltic concrete surface layer of thickness D is shown to be subjected to a tensile stress distribution with depth, as previously discussed. It is assumed that within the surface layer there are potential crack sites uniformly spaced at a distance S. At each of these crack sites the induced thermal stresses can potentially cause a crack to propagate through the surface layer (figure 2), at which time it is assumed that a transverse crack will be visible on the pavement surface. It is assumed that each of these cracks can propagate at different rates due to spatial variation of the relevant material properties within the surface layer.

The thermal cracking model consists of two primary parts:

- A mechanics-based model that calculates the progression of a vertical crack at one crack site having average material properties.

- A probabilistic model that calculates the global amount of thermal cracking visible on the pavement surface from the current average crack depth and the assumed distribution of crack depths within the surface layer.

Model Components

Flow diagrams of the thermal cracking model are shown in figures 3 and 4. Figure 3 illustrates the interrelationships between the five major components of the model. These five components are the:

- Inputs module
- Indirect tensile test at low temperatures (ITLT) and ITLT transformation model
- Environmental effects model
- Pavement response model
- Pavement distress model

Figure 4 provides more detailed information for each of the model components. Hiltunen and Roque [2] have described each of the model components in detail. The

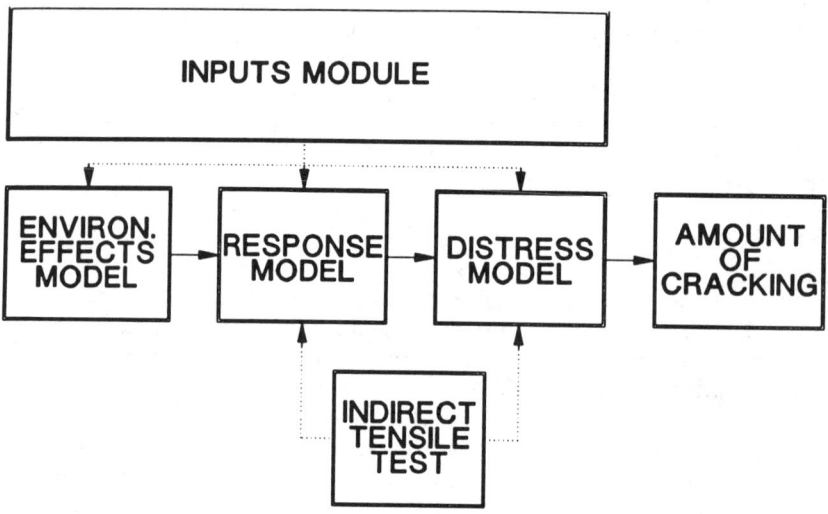

Figure 3. Major Components of the Thermal Cracking Model

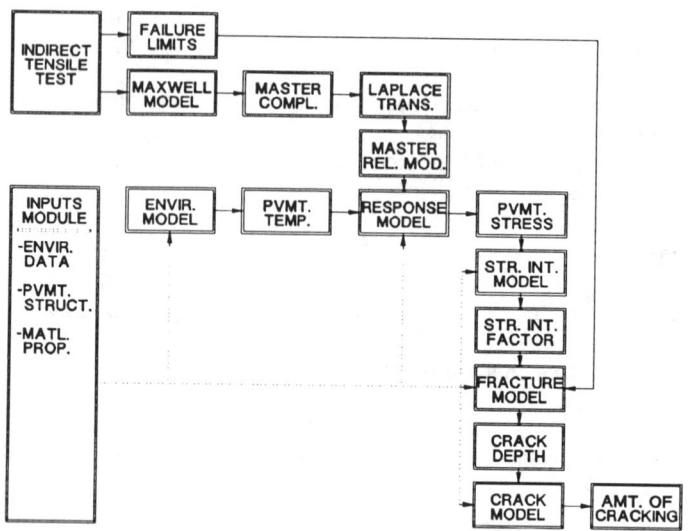

Figure 4. Detailed Schematic of the Thermal Cracking Model

following sections will describe the materials model used to determine viscoelastic material properties from ITLT test data.

INDIRECT TENSILE TEST AT LOW TEMPERATURES (ITLT)

If one considers the primary mechanism of thermal cracking shown in figure 1 (thermal stress development and crack propagation), the primary material properties controlling this mechanism are the viscoelastic properties, which control thermal stress development, and the fracture properties, which control the rate of crack development. These are the properties that are measured and controlled by the new SHRP mixture specification test for thermal cracking (figure 5). The test is identified as the indirect tensile test at low temperatures (ITLT).

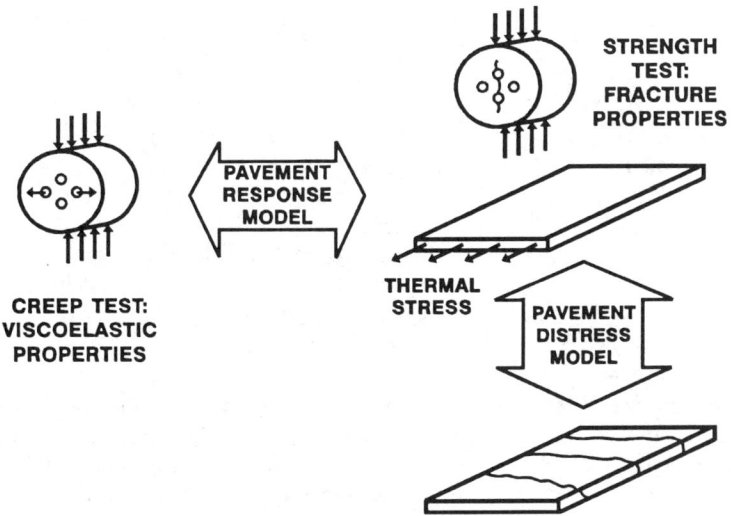

Figure 5. Materials Characterization with the ITLT

The ITLT is conducted on asphaltic concrete specimens in two phases:

- Short-term creep tests (approximately 1000 seconds) at multiple temperatures are used to determine the viscoelastic properties.

- A tensile strength test is used along with the viscoelastic properties to determine the fracture properties of the mixture.

Example creep compliance curves measured at -20, -10, and 0 °C on two faces of three specimens each are shown in figures 6-8. The results forwarded to the materials model are the average 1000-second creep compliance curves determined from the multiple specimens and faces at three temperatures (see figure 9) and the average tensile strength at -10 °C. The average data are typically based upon tests on three replicate specimens, as shown in figures 6-8.

VISCOELASTIC MATERIAL PROPERTY RELATIONSHIPS

The viscoelastic properties of the asphaltic concrete mixture control the level of stress development during cooling. More specifically, the time- and temperature-dependent relaxation modulus of the mixture is the property needed to compute thermal stresses in the pavement according to the following constitutive equation [2]:

$$\sigma(\xi) = \int_0^\xi E(\xi-\xi')\frac{d\epsilon}{d\xi'}d\xi' \qquad (1)$$

where:
$\sigma(\xi)$ = stress at reduced time ξ
$E(\xi-\xi')$ = relaxation modulus at reduced time $\xi-\xi'$
ϵ = strain at reduced time ξ $(= \alpha(T(\xi') - T_0))$
α = linear coefficient of thermal contraction
$T(\xi')$ = pavement temperature at reduced time ξ'
T_0 = pavement temperature when $\sigma = 0$
ξ' = variable of integration

A generalized Maxwell model was selected to represent the viscoelastic properties of the asphaltic concrete mixture in relaxation. A schematic representation of the model is shown in figure 10. Mathematically, the relaxation modulus for a generalized Maxwell model can be expressed according to the following Prony series:

$$E(\xi) = \sum_{i=1}^{N+1} E_i e^{-\xi/\lambda_i} \qquad (2)$$

where:
$E(\xi)$ = relaxation modulus at reduced time ξ
E_i, λ_i = Prony series parameters for master relaxation modulus curve (spring constants or moduli and relaxation times, respectively, for the Maxwell elements)

This function describes the relaxation modulus as a function of time at a single temperature, which is generally referred to as the reference temperature. The

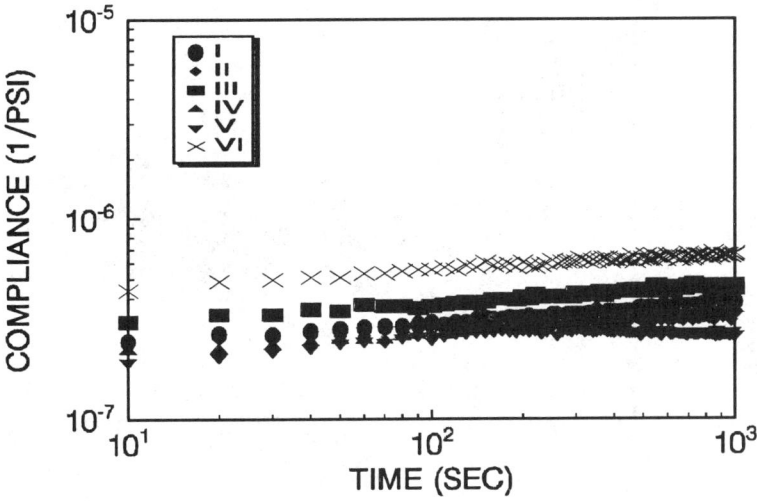

Figure 6. Creep Compliance Curves at -20 °C for PTI Section 38 (1 psi = 6.89 kPa)

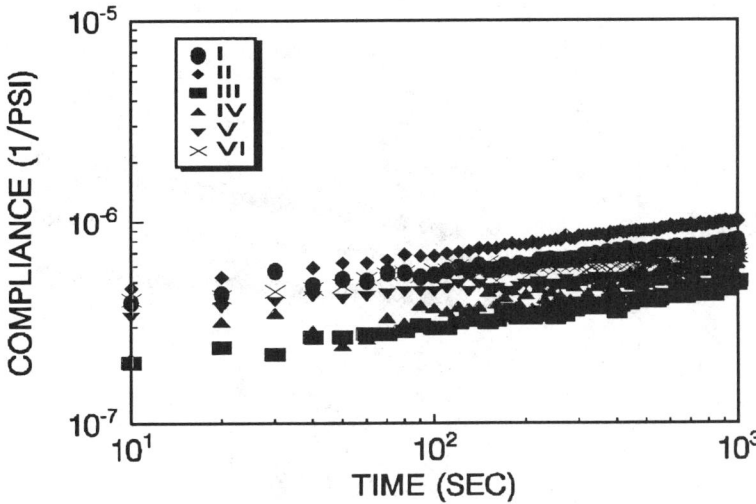

Figure 7. Creep Compliance Curves at -10 °C for PTI Section 38 (1 psi = 6.89 kPa)

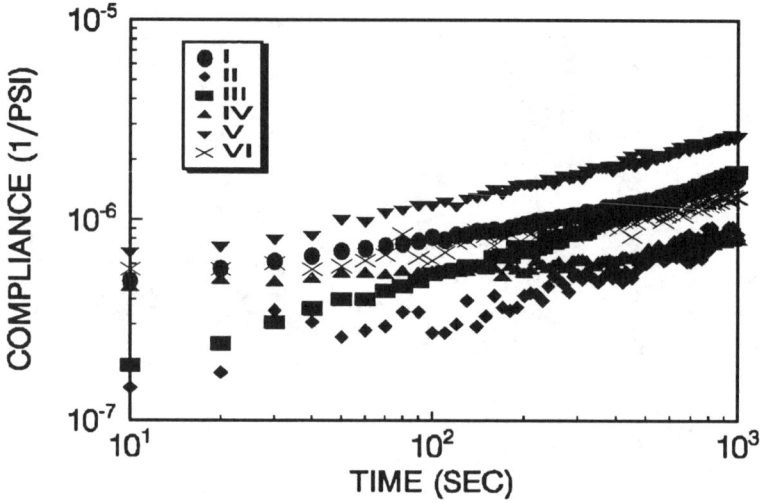

Figure 8. Creep Compliance Curves at 0 °C for PTI Section 38 (1 psi = 6.89 kPa)

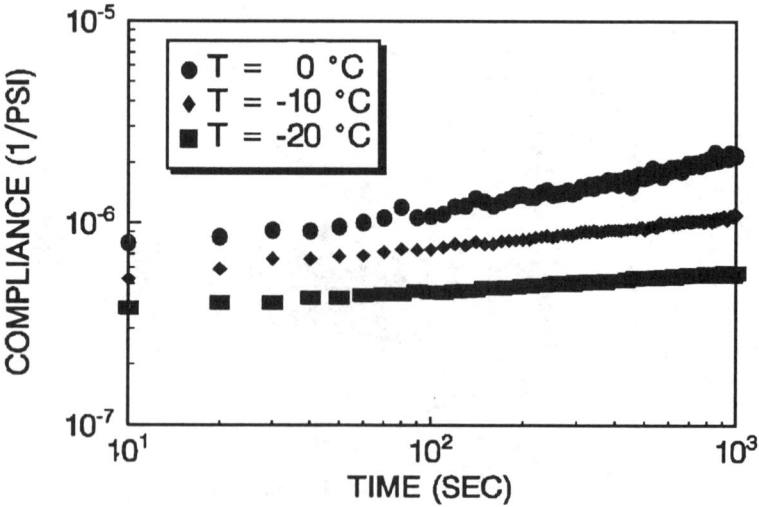

Figure 9. Average Creep Compliance Curves for PTI Section 38 (1 psi = 6.89 kPa)

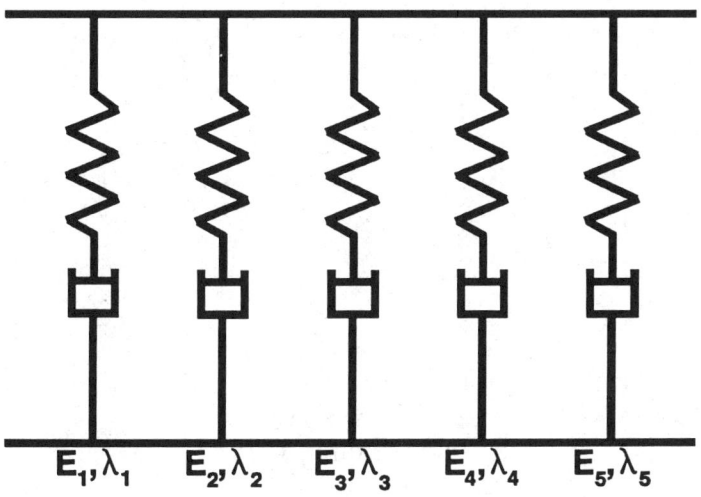

Figure 10. Generalized Maxwell Model for Relaxation

function defined at the reference temperature is called the master relaxation modulus curve. Relaxation moduli at other temperatures are determined by using the method of reduced variables (time-temperature superposition), which means that the mixture is assumed to behave as a thermorheologically simple material. Relaxation moduli at other temperatures are determined by replacing real time (i.e., time corresponding to the temperature of interest) with reduced time (i.e., time corresponding to the temperature at which the relaxation modulus is defined) according to the following equation:

$$\xi = \frac{t}{a_T} \qquad (3)$$

where: ξ = reduced time
        t = real time
        a_T = temperature shift factor

The relaxation modulus function is obtained by transforming the following time-dependent creep compliance function:

84   ENGINEERING PROPERTIES OF ASPHALT MIXTURES

$$D(\xi) = D(\infty) - \sum_{i=1}^{N} D_i e^{-\xi/\tau_i} + \frac{\xi}{\eta_v} \tag{4}$$

or

$$D(\xi) = D(0) + \sum_{i=1}^{N} D_i(1-e^{-\xi/\tau_i}) + \frac{\xi}{\eta_v} \tag{5}$$

where:

$$D(\infty) = D(0) + \sum_{i=1}^{N} D_i \tag{6}$$

and:
$D(\xi)$ = creep compliance at reduced time ξ
ξ = reduced time = t/a_T
a_T = temperature shift factor
$D(\infty), D(0), D_i, \tau_i, \eta_v$ = Prony series parameters

Prony series parameters and shift factors are obtained by performing creep compliance tests at multiple temperatures (phase one of the ITLT) and mathematically shifting data from different temperatures to establish one smooth, continuous curve. This process is illustrated conceptually in figure 11. The resulting curve is called the master creep compliance curve.

ITLT TRANSFORMATION MODEL

The purpose of the ITLT transformation model is to determine the master relaxation modulus curve from the creep compliance measurements. The transformation is accomplished in two steps. First, the master creep compliance curve is generated from the creep compliance test results at different temperatures. Second, the master relaxation modulus curve is determined from the master creep compliance curve via the Volterra integral.

Master Creep Compliance Curve

A nonlinear regression routine is used to determine the master creep compliance curve from the creep compliance curves determined at multiple temperatures. The regression is performed in two steps. First, a regression is performed to simultaneously determine the temperature shift factors (a_T) and the

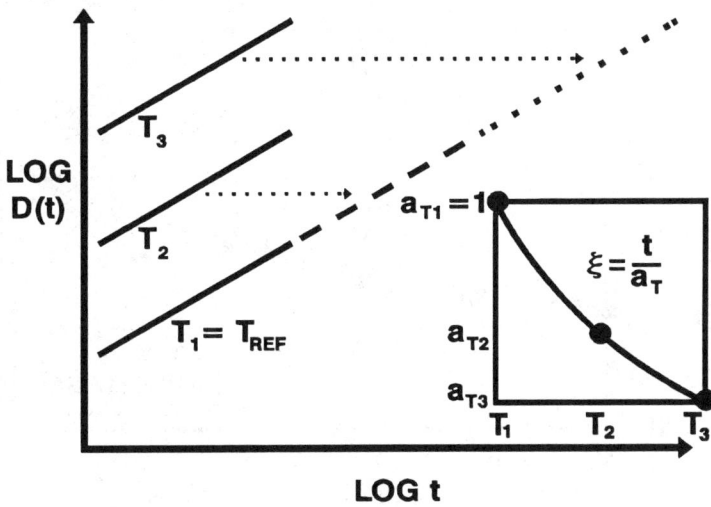

Figure 11. Development of the Master Creep Compliance Curve

parameters for the Prony series representation of the master creep compliance curve. The function for the regression is obtained by substituting equation 3 into equation 5. One of the test temperatures is selected as the reference temperature for the master curve, and thus the creep compliance curve at this temperature is fixed in time ($a_T=1$). The regression determines the amount of time (horizontal) shift required for the curves at the remaining temperatures to result in a smooth master curve. Each of these remaining creep compliance curves will thus have a shift factor (a_T) associated with it.

In conjunction with the determination of the shift factors, the regression determines the coefficients for the Prony series. Four terms of the Prony series ($N=4$) have been found to be sufficient to fit the data accurately when 1000-second creep compliance curves at three temperatures (-20, -10, and 0 °C) are used to construct the master curve. The master creep compliance curve and the shift factors as a function of temperature for the test results previously presented in figure 9 are shown in figures 12 and 13, respectively.

As shown in figure 12, the data from figure 9 shift horizontally to form a smooth master curve, indicating that the principle of time-temperature superposition is applicable to this asphaltic concrete mixture. Figure 12 also clearly shows that the Prony series fits the master curve data well, suggesting that a generalized Maxwell model adequately represents the creep response of asphaltic concrete mixtures at low temperatures. Finally, figure 13 shows that the shift factor versus temperature

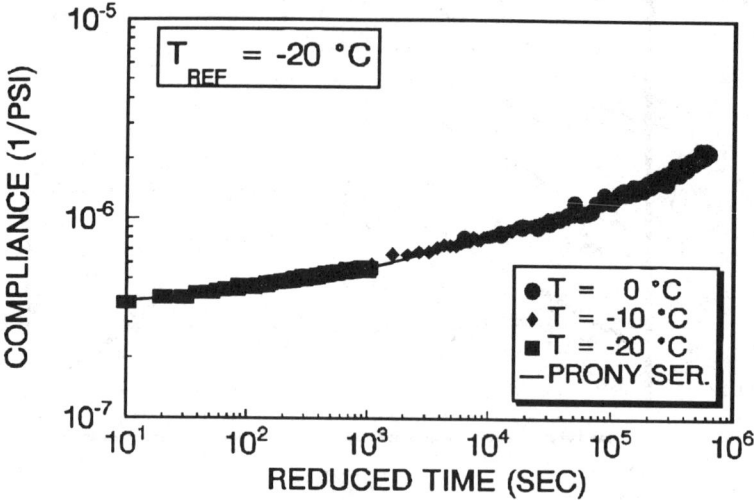

Figure 12. Master Creep Compliance Curve for PTI Section 38 (1 psi = 6.89 kPa)

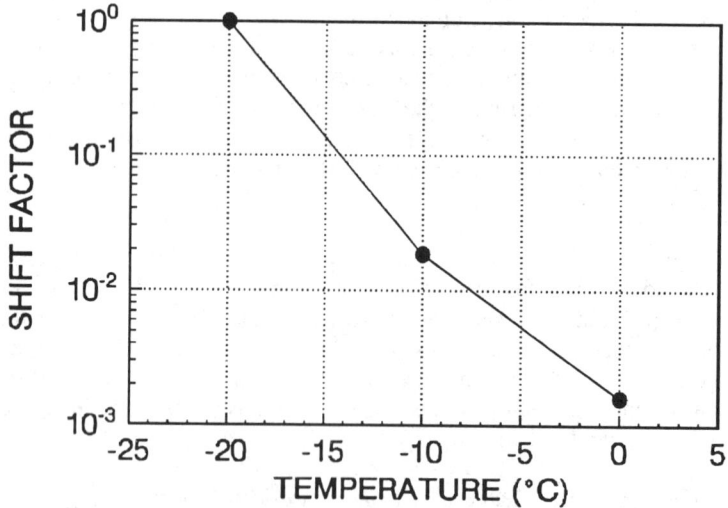

Figure 13. Shift Factor versus Temperature for PTI Section 38

relationship for asphaltic concrete mixtures at low temperatures is not linear on a semi-logarithmic plot. Other thermorheologically simple materials, including asphalt cement binders at low temperatures, are often found to have a linear relationship between the logarithm of the shift factor and temperature. In the authors' experience with over 40 asphaltic concrete mixtures, the above observations are typical. In fact, the nonlinearity of the shift factor relationship is often more pronounced than as shown in figure 13 for PTI section 38.

In addition to fitting the master curve data well, the Prony series was chosen to represent the master creep compliance curve for two important reasons:

- The Prony series greatly simplifies the transformation of the master creep compliance curve to the master relaxation modulus curve (described in the next section).

- The Prony series greatly simplifies the solution of the viscoelastic constitutive model used to calculate pavement stresses [2].

Because of the nonlinear behavior noted above, the shift factor-temperature relationship is modeled within the thermal cracking model as piecewise linear between shift factors determined at the specified test temperatures, assuming a semi-log relationship between the natural logarithm of the shift factors versus arithmetic temperature. In other words, shift factors at arbitrary temperatures are determined by linearly interpolating, assuming a semi-log relationship, between the shift factors determined from the regression. Linear extrapolation, assuming a semi-log relationship, is performed to obtain shift factors at temperatures outside the range of measurements.

The second step in the regression routine is to fit a second functional form to the master creep compliance information. This second functional form is the following power model:

$$D(\xi) = D_0 + D_1 \xi^m \tag{7}$$

where $D(\xi)$ and ξ are as defined previously, and D_0, D_1, and m are the coefficients of the functional form. The primary purpose for fitting this functional form is to determine the parameter m. This parameter is essentially the slope of the linear portion of the master creep compliance curve on a log-log plot (figure 14). As discussed by Roque, Hiltunen, and Stoffels [1] and Hiltunen and Roque [2], m has been found to be an important parameter in distinguishing between the thermal cracking performance of different materials, and is a direct input into the crack depth (fracture) portion of the thermal cracking model.

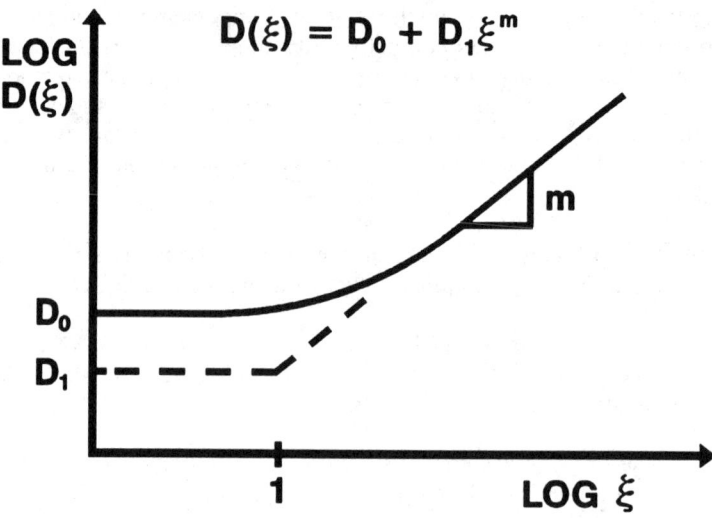

Figure 14. Power Model for Master Creep Compliance Curve

Master Relaxation Modulus Curve

As previously described, the viscoelastic constitutive equation used in the pavement response model requires that the time and temperature dependent relaxation modulus of the material be known. It is common to formulate the constitutive equation in terms of relaxation modulus when the stress response to a known strain input is desired, which is the case here (equation 1). However, it is also accepted and understood that creep tests on viscoelastic materials are typically easier to conduct and the results are more reliable than relaxation test results. Therefore, the creep phase of the ITLT was developed for determining the viscoelastic properties.

The viscoelastic property determined from a creep test is known as the creep compliance. The creep compliance is simply the time dependent strain divided by the constant stress. However, the property required for the stress predictions is the relaxation modulus, as discussed above. Sometimes the relaxation modulus is approximated as simply the inverse of the creep compliance. However, this is not correct. The inverse of the creep compliance is the creep modulus (or creep stiffness), and not the relaxation modulus. Although under some conditions (e.g., low temperatures and short loading times with hard materials) the two moduli are approximately equal, this is generally not the case. Since the creep compliance and the true relaxation modulus are related, it is relatively simple to determine the true relaxation modulus, rather than approximate it. As previously discussed, the calculations are particularly easy if a Prony series is used to represent the master creep compliance curve.

For a viscoelastic material, the relationship between creep compliance and relaxation modulus is given by the Volterra integral:

$$\int_0^\infty D(t-\tau)\frac{dE(\tau)}{d\tau}d\tau = 1 \tag{8}$$

Taking the Laplace transformation of each side results in:

$$L[D(t)] * L[E(t)] = \frac{1}{s^2} \tag{9}$$

where:
- $L[D(t)]$ = the Laplace transformation of the creep compliance, $D(t)$
- $L[E(t)]$ = the Laplace transformation of the relaxation modulus, $E(t)$
- s = the Laplace parameter
- t = time (or reduced time, ξ)

A computer program has been developed and included within the thermal cracking model to solve this equation for the master relaxation modulus, $E(\xi)$, given the master creep compliance, $D(\xi)$. The program essentially performs the following steps:

1. Computes the Laplace transformation of the master creep compliance, $L[D(\xi)]$, where $D(\xi)$ is defined by the Prony series described in equation 4

2. Multiplies $L[D(\xi)]$ by s^2

3. Computes the reciprocal of $s^2 * L[D(\xi)]$, which is $L[E(\xi)]$

4. Computes $E(\xi)$, which is the inverse Laplace transformation of the step 3 result. $E(\xi)$ will then have the Prony series functional form given in equation 2.

The master relaxation modulus curve determined from the master creep compliance curve presented in figure 12 is shown in figure 15.

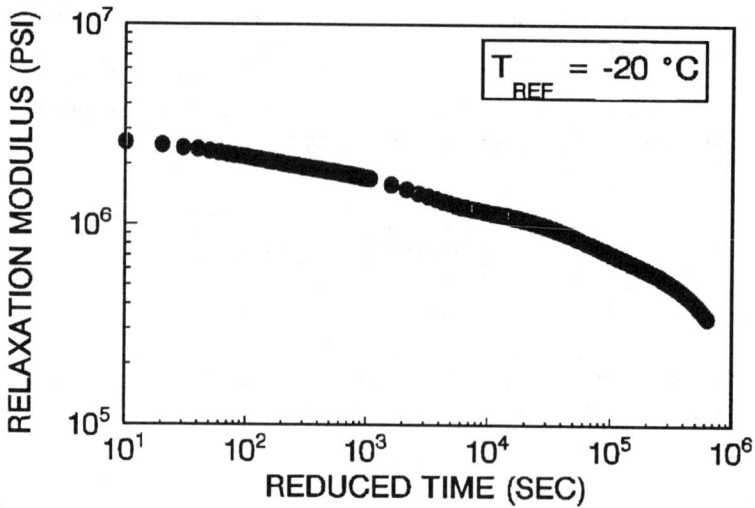

Figure 15. Master Relaxation Modulus Curve for PTI Section 38 (1 psi = 6.89 kPa)

The transformation program has been independently verified by comparing the results to those produced by *Mathematica*, which is a computer software system that is able to perform Laplace transformations [6]. An example from these comparisons is shown in Table 1 for PTI section 38. As observed, the Prony series parameters from the two independent determinations are virtually identical.

Also, comparisons made between the relaxation modulus computed as described and the creep modulus have shown that they in fact are not equivalent, and thus it would not be accurate to approximate the relaxation modulus with the creep modulus. In all cases compared the differences were not dramatic; however, in all cases differences did exist.

Summary

In summary, the following procedure is followed in determining the master relaxation modulus function from ITLT test results:

1. Perform 1000-second creep tests at multiple temperatures and reduce the data (i.e., determine creep compliance versus time) following the ITLT procedures described by Roque and Buttlar [3,4].

2. Determine the master creep compliance curve and the temperature shift factors from the 1000-second creep compliance curves at multiple

Table 1. Master Relaxation Modulus Prony Series Parameters for PTI Section 38

Method	Moduli (psi)					Relaxation Times (sec)				
	E_1	E_2	E_3	E_4	E_5	λ_1	λ_2	λ_3	λ_4	λ_5
TCMODEL	200788.	481712.	746006.	510000.	756131.	24.5578	117.270	2097.13	44221.3	792186.
Mathematica	200788.	481712.	746007.	509999.	756131.	24.5578	117.270	2097.14	44221.4	792186.

Notes: TCMODEL = The Penn State Thermal Cracking Model developed as part of SHRP A-005

1 psi = 6.89 kPa

temperatures. Fit both the Prony series and the power model functions to the master creep compliance curve.

3. Determine the master relaxation modulus curve from the master creep compliance curve via the Volterra integral and the Prony series fit to the master creep compliance curve. The temperature shift factors for the master relaxation modulus curve are as determined for the creep compliance data, i.e., the same shift factors are applicable to both the creep and relaxation data.

CONCLUSIONS

Based upon the data presented and discussed herein, the following conclusions are appropriate:

- The principle of time-temperature superposition can adequately represent the time- and temperature-dependent stiffness properties of asphaltic concrete mixtures at low temperatures.

- A generalized Maxwell model (Prony series) can accurately represent the master creep compliance curve of asphaltic concrete mixtures at low temperatures.

- The shift factor versus temperature relationship for asphaltic concrete mixtures at low temperatures is typically nonlinear on a semi-logarithmic scale.

ACKNOWLEDGMENTS

The work described in this publication was the result of the efforts of a team of faculty and student researchers at The Pennsylvania State University. The authors gratefully acknowledge the efforts of the following individuals who made the successful completion of this work possible: Dr. Steven F. Arnold, Dr. M. G. Sharma, Dr. Shelley M. Stoffels, William G. Buttlar, Tariq Farwana, Namho Kim, Kevin Knechtel, Wendy Lauritzen, Srinivas Reddy, Pedro Romero, Nader Tabatabaee, and Vivek Tandon. A comprehensive description of the work performed by this team can be found in the final report by Lytton et al. [7]. Financial support for this work was provided by the Strategic Highway Research Program (SHRP) under research contract A-005. The authors gratefully acknowledge this support.

REFERENCES

[1] Roque, R., Hiltunen, D. R., and Stoffels, S. M. (1993), "Field Validation of SHRP Asphalt Binder and Mixture Specification Tests to Control Thermal Cracking Through Performance Modeling," *Journal of the Association of Asphalt Paving Technologists*, Vol. 62, pp. 585-600.

[2] Hiltunen, D. R. and Roque, R. (1994), "A Mechanics-Based Prediction Model for Thermal Cracking of Asphaltic Concrete Pavements," *Journal of the Association of Asphalt Paving Technologists*, Vol. 63.

[3] Roque, R. and Buttlar, W. G. (1992), "The Development of a Measurement and Analysis System to Accurately Determine Asphalt Concrete Properties Using the Indirect Tensile Mode," *Journal of the Association of Asphalt Paving Technologists*, Vol. 61, pp. 304-332.

[4] Buttlar, W. G. and Roque, R. (1994), "Experimental Development and Evaluation of the New SHRP Measurement and Analysis System for Indirect Tensile Testing of Asphalt Mixtures at Low Temperatures," *Research Record No. 1454*, Transportation Research Board, Washington, D. C., pp. 163-171.

[5] Haas, R., Meyer, F., Assaf, G., and Lee, H. (1987), "A Comprehensive Study of Cold Climate Airport Pavement Cracking," *Proceedings of the Association of Asphalt Paving Technologists*, Vol. 56, pp. 198-245.

[6] Wolfram, S. (1991), *Mathematica: A System for Doing Mathematics by Computer*, 2nd ed., Addison-Wesley Publishing Company, Inc., Reading, Massachusetts, 961 pp.

[7] Lytton, R. L., Roque, R., Uzan, J., Hiltunen, D. R., Fernando, E., and Stoffels, S. M. (1993), "Performance Models and Validation of Test Results," *Final Report to Strategic Highway Research Program; Asphalt Project A-005*, July, 529 pp.

Relating Material Properties to Permanent Deformation

T. F. Fwa,[1] B. H. Low,[2] and S. A. Tan[3]

Behavior Analysis of Asphalt Mixtures
Using Triaxial Test-Determined Properties

REFERENCE: Fwa, T. F., Low, B. H., and Tan, S. A., "**Behavior Analysis of Asphalt Mixtures Using Triaxial Test-Determined Properties,**" Engineering Properties of Asphalt Mixtures and the Relationship to their Performance, ASTM STP 1265, Gerald A. Huber and Dale S. Decker, Eds., American Society for Testing and Materials, Philadelphia, 1995.

ABSTRACT: The mix design procedure for asphalt mixtures by the Marshall method does not furnish material property parameters that can be used analytically for structural thickness design and performance analysis of pavements. This research illustrates that this missing link can be bridged by the use of laboratory triaxial test-determined properties to characterize asphalt mixtures. By means of finite element analysis using triaxial test-determined properties as input, it is illustrated that the stress-strain behavior of asphalt specimens up to failure can be simulated for the following three loading conditions: triaxial test, Marshall test, and indirect tension test. Laboratory tests involving four different asphalt mixtures are performed to provide the verification data.

KEYWORDS: triaxial test, Marshall test, indirect tension test, asphalt mixtures, finite element analysis, cohesion, angle of friction, elastic modulus

INTRODUCTION

The Marshall stability test is used worldwide by pavement engineers as a basis of mix design for asphaltic paving materials. The major limitations of the test arise mainly from the fact that it is empirically based. The results of the test cannot be used directly in pavement thickness design. It is not possible to use the test results to predict or study analytically the behavior of the design mixture under traffic loading.

---

[1]Assoc. Professor, [2]Research Assistant, [3]Senior Lecturer, Department of Civil Engineering, National University of Singapore, 10 Kent Ridge Crescent, Singapore 0511.

Modern day pavement evaluation and design involve analysis of pavement response under traffic loading and environmental forces. To this end, a number of theoretical models have been proposed by researchers since early 1950s. One of the earliest models studied was based on the Mohr-Coulomb theory involving material properties of cohesion, c, and angle of friction, ϕ [1-4]. However, there has been little progress in the application of triaxial test-determined properties in either mix design or pavement response analysis.

The advantage of using triaxial test is that it is a well established test in civil engineering, and it provides fundamental engineering properties which can be used for analyzing the behavior of materials under loads. This study demonstrates using laboratory test results that this is also applicable in the analysis of asphalt mixtures. The analysis is performed using the finite element method that adopts the Drucker-Prager failure criterion [5, 6] for modeling the behavior of asphalt mixtures. The usefulness of the analysis is illustrated by applying it to predict the stress-strain behavior of asphalt specimens under three different laboratory loading conditions.

SCOPE OF STUDY

The main focus of the study was to examine the applicability of the proposed approach for analyzing the stress-strain behavior of asphalt mixtures under applied loads. This is achieved by applying triaxial test-determined properties of asphalt mixtures in the analysis of the mixtures' behavior under three different laboratory test loading conditions, and comparing the predicted behavior with measured responses. The three loading conditions studied were: (a) triaxial test, (b) Marshall test, and (c) indirect tension test.

The major elements of work involved the design and installation of triaxial test equipment, and finite element analyses of asphalt mixture behavior using the triaxial test-determined properties. Four different asphalt mixture types were tested to provide the verification data. They included two dense graded mixes, an open graded mix and a sand asphalt. The idea of including the different mix types was to illustrate the ability of the described approach to analyze mixture behavior regardless of the type of mix.

DETERMINATION OF ASPHALT MIXTURE PROPERTIES BY TRIAXIAL TEST

Triaxial Test Setup

The cell used in the present study is a conventional 6-inch (152 mm) diameter triaxial cell for soil testing, but modified with fittings to provide temperature control required for asphalt mixture testing. The entire triaxial cell is immersed in a perspex water bath set at 2 to 3°C above the test temperature. A thermal reservoir is employed to provided heated water at the desired test temperature to the cell. The time required to heat a specimen to uniform test temperature varies from about 20 min. for test temperature of 40°C to about 60 min. for test temperature of 60°C.

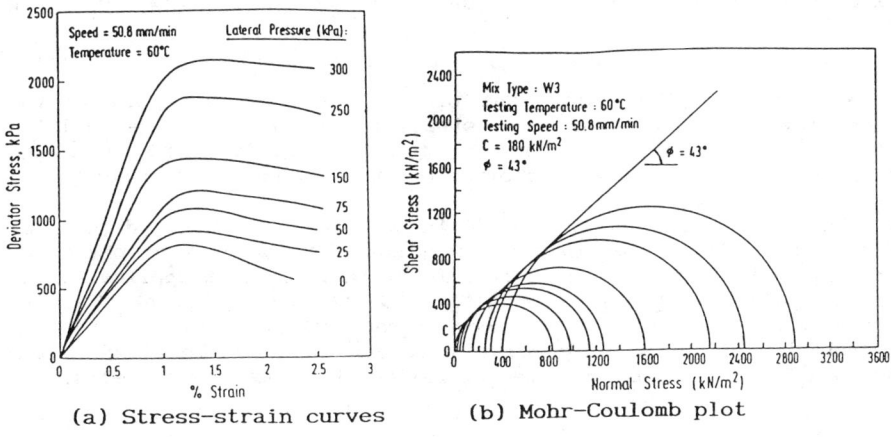

Fig. 1—Determination of asphalt mixture properties by triaxial tests

Each triaxial test specimen measures 102 mm (4 in.) in diameter and about 200 mm (approximately 8 in.) in height. It is placed between the loading platens and sealed from water by means of an impervious rubber membrane. Each test is carried out at constant temperature, strain rate, and confining water pressure. The load and vertical displacement are recorded continuously.

Computation of Asphalt Mixture Properties

Tests at four or more confining pressures ranging from 0 to 400 kPa (58 psi) are performed to provide the necessary data for determining the cohesion c and friction angle ϕ of an asphalt mixture at a given test temperature and strain rate. Three specimens were tested for each confining pressure. Figure 1(a) shows typical stress-strain curves obtained from the triaxial test. It is observed that the curves exhibit essentially linear response up to a strain of about 1.0%. The slope of each line defines the elastic modulus E of the mixture for the test conditions.

The peak stresses of the stress-strain curves of Fig. 1(a) were determined to plot Mohr circles as shown in Fig. 1(b). Each of the Mohr circles shown is the average of the peak stress of three specimens tested at the same confining pressure. A linear failure line was fitted to obtain the cohesion c and friction angle ϕ.

BEHAVIOR MODELING OF ASPHALT MIXTURES

Drucker-Prager Failure Criterion

In this study the stress-strain behavior of asphalt mixtures is modeled by an elasto-plastic idealization with the Drucker-Prager failure criterion [5, 6]. The Drucker-Prager criterion defines a yield function f as

$$f = \sqrt{J_{2D}} - \alpha J_1 - k \tag{1}$$

where J_1 is the first invariant of the stress tensor, J_{2D} is the second invariant of the deviator stress tensor, α and k are positive material parameters. α and k can be determined from the slope and intercept of failure envelop plotted on the J_1 versus $\sqrt{J_{2D}}$ space.

The above Drucker-Prager yield function can be expressed in terms of triaxial test-determined properties of asphalt materials by matching with the Mohr-Coulomb failure criterion. Such application of Mohr-Coulomb theory in conjunction with Drucker-Prager failure criterion has been used by a number of researchers [7-9] for modeling the behavior of soils and granular materials. It can be shown that α and k in Eq. (1) can be computed as follows for compression tests,

$$\alpha = \frac{2 \sin\phi}{\sqrt{3} \ (3-\sin\phi)} \tag{2}$$

$$k = \frac{6c \cdot \cos\phi}{\sqrt{3} \ (3-\sin\phi)} \tag{3}$$

The input parameters to an analysis based on Drucker-Prager model are thus fully defined by c and ϕ through the use of Eqs. (2) and (3).

Finite Element Modeling Incorporating Drucker-Prager Criterion

The use of Drucker-Prager model makes it convenient to analyze stress-strain behavior of asphalt mixtures by means of finite element formulation. The following four material parameters are required as input: cohesion c, angle of friction ϕ, elastic modulus E, and Poisson ratio ν. The first three parameters were determined from triaxial tests. Values of ν in the range of 0.35 to 0.5 were used in trial analysis and found to have negligible effects on the results of simulation. 0.45 was used for ν in this study.

Finite Element Model for Triaxial Test

The triaxial test on a cylindrical specimen is axisymmetric in both geometry and loading pattern, and only one quarter of the test specimen needs to be modeled. The finite element mesh used is shown in Fig. 2. There is a total of 72 eight-noded elements, including 12 for the brass loading platen. The brass platen was assumed elastic with modulus of 100 GPa (14.5 x 10^6 psi) and Poisson ratio of 0.35. The constant confining pressure was represented by nodal point loads along the side of the specimen. Incremental loads were applied until the specimen failed. Load increments of 0.1 kN was used when the failure load was approached. Failure was assumed to occur when one of the element of the finite element mesh became plastic.

Finite Element Model for Marshall Test

Due to symmetry in the Marshall test layout, it is necessary to model only one-quarter of the specimen. Figure 3 shows the finite element mesh that also contains one-half of the upper test head. The

FWA ET AL. ON TEST-DETERMINED PROPERTIES 101

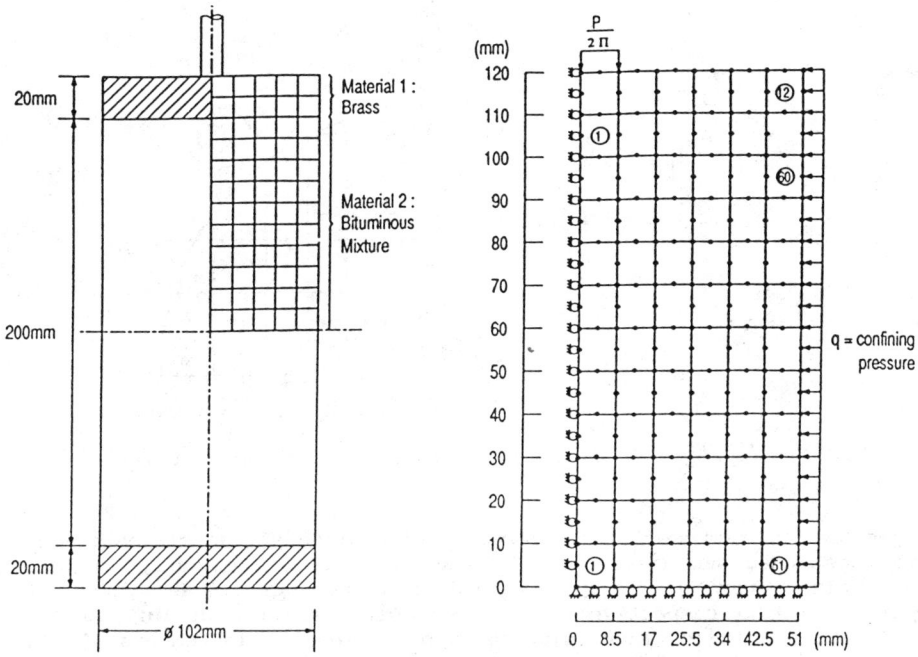

Fig. 2—Finite element mesh for modeling triaxial test

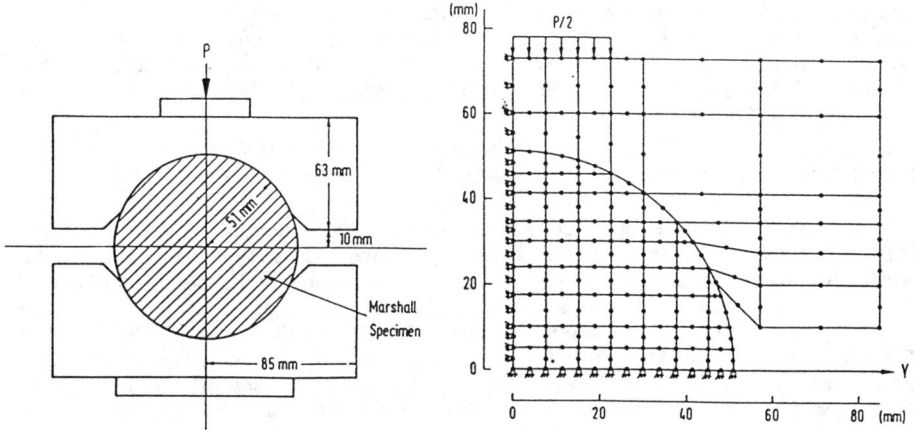

Fig. 3—Finite element mesh for modeling Marshall test

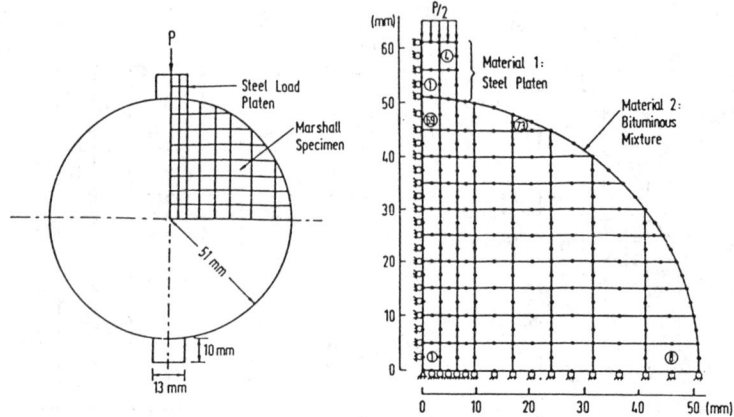

Fig. 4—Finite element mesh for modeling indirect tension test

specimen was represented by 52 rectangular plane stress elements while the test head was discretized into 20 elements. The Young's modulus and Poisson ratio of the steel head were taken as 207 GPa (30×10^6 psi) and 0.35 respectively. The specimen behavior during Marshall test was studied by increasing the applied load at increments of 1 kN initially, and 0.1 kN at the final stages of the test. No slip was assumed at the interface between test heads and the specimen.

Finite Element Model for Indirect Tension Test

As in the case for Marshall test, only one quarter of the test specimen needs to be considered in modeling the indirect tension test. Fig. 4 shows the finite element mesh with the specimen and loading strip represented by 73 and 4 rectangular plane stress elements respectively. The steel loading was assumed elastic with Young's modulus of 207 GPa (30×10^6 psi) and Poisson ration of 0.35. The same load increment scheme used for Marshall test simulation was also adopted in this case.

EXPERIMENTAL TEST PROGRAM

In view of the temperature and loading rate dependency of asphalt mixture behavior, the three loading tests, namely triaxial test, Marshall compression test, and indirect tension test, were performed at common constant temperatures and loading rates to facilitate comparison. Five test temperatures of $40°C$, $45°C$, $50°C$, $55°C$, and $60°C$, and five loading rates of 4, 20, 40, 50.8 and 60 mm/min were used. The range of test temperature provides a reasonable coverage of the service temperature of asphaltic pavements in the hot tropical climate of Singapore. Except for the choices of test temperature and loading rate, the Marshall and indirect tension test procedures followed those described in ASTM Test Method for Resistance to Plastic Flow of Bituminous Mixtures using Marshall Apparatus (D 1559) and ASTM

Test Method for Indirect Tension Test for Resilient Modulus of Bituminous Mixtures (D 4123) respectively.

Materials for Asphalt Mixtures

Four asphalt mixtures, designated as W3, W1, B1, and SA, were tested. The mix proportions and aggregate gradations for the four mixtures are shown in Table 1. W1 and W3 are dense-graded mixtures used for wearing course. B1 is an open-graded mixture, and SA is a sand asphalt mixture. Asphalt cement of 60/70 penetration grade and granite aggregates, the only type of aggregate available in Singapore, were used for preparing W1, W3 and B1 mixtures. The same asphalt cement was also used for SA mixtures. The binder content of each mixture was the optimum asphalt content by weight of total mix determined using the Marshall method of mix design [10]. The specimens of W1 mix had a mean bulk specific gravity of 2.317 with a standard deviation of 0.013, and a mean percent air void of 4.91% with a standard deviation of 0.54%. The corresponding data of specific bulk gravity are 2.355 and 0.008 for W3 specimens, 2.406 and 0.003 for B1 specimens, and 2.066 and 0.027 for SA specimens. The corresponding data of air void are 4.49% and 0.33% for W3 specimens, 3.58% and 0.12% for B1 specimens, and 6.54% and 0.24% for SA specimens.

Table 1 — Mix Proportions and Aggregate Gradations

Mix Type	Binder Content	Maximum Theoretical Density g/cc	Aggregate size distribution (% passing)								
			25 mm	19 mm	13 mm	9.5 mm	6.4 mm	3.2 mm	1.2 mm	0.3 mm	0.075 mm
W1	5.5%	2.427	100	100	100	100	95	74	47	27	6
W3	5.5%	2.469	100	100	95	87	77	58	37	19	6
B1	5.0%	2.495	100	88	74	64	55	43	28	15	5.5
SA	9.0%	2.324	100	100	100	100	100	100	93	60	15

Specimen Preparation

The Marshall and the indirect tension tests were performed on 4-in.(102 mm) diameter cylindrical specimen with a height of about 2.5 in. (64 mm). The specimens for triaxial tests had the same diameter but measured about 200 mm (approximately 8 in.) in height. Since the main objective of the present study was to examine specimen behavior during Marshall and indirect tension tests using triaxial test-determined properties, it was desirable to have all specimens compacted to about the same density by a similar compaction procedure.

A special study was conducted to determine the compaction procedures that would produce 64 mm and 200 mm tall specimens of about the same density. The details of the study are reported elsewhere [11]. It was found that single-layer double plunger compression would produce uniform density in specimens of the two sizes. For 64 mm tall specimens, a static pressure of 25.kPa applying for 5 minutes achieved a uniform density distribution closely similar to those compacted by the standard Marshall method of drop-hammer compaction. In the case of 200 mm tall cylindrical triaxial specimens, maintaining the same static pressure for 20 minutes produced specimens of about the same

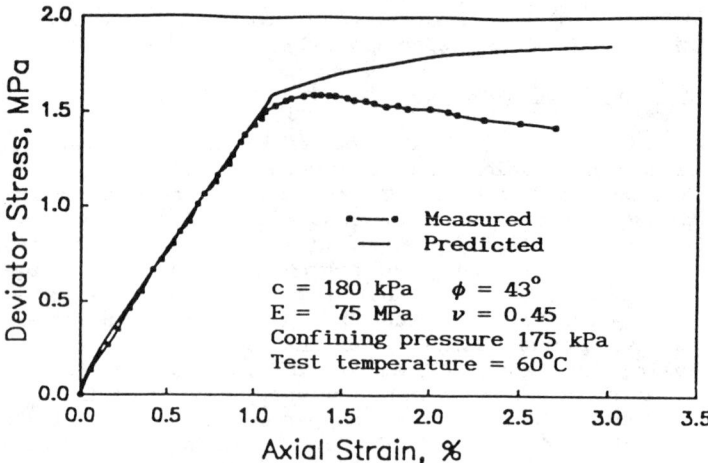

Fig. 5—Predicted and measured stress-strain behavior of asphalt specimen in triaxial test

density as that of 64 mm tall specimens. As a verification, Marshall tests and indirect tension tests were also performed on specimens cut from 200 mm tall triaxial specimens. Test results were found to be statistically not different from those obtained from 64 mm tall specimens prepared by the conventional drop hammer compaction method. The double-plunger compression method was therefore adopted for compaction of all test specimens.

COMPARISON OF PREDICTED AND MEASURED BEHAVIOR

Behavior of Specimens in Triaxial Tests

Shown in Fig. 5 is an example that compares finite element predicted response and measured stress-strain behavior of specimens in triaxial tests. The elastic modulus E was determined from triaxial tests performed at the particular confining pressure. The predicted response described fairly well the stress-strain behavior up to the peak stress. Discrepancies are observed after the peak stress because post-failure behavior cannot be modeled by the finite element analysis.

The overall statistical correlation coefficient r between the predicted and measured maximum deviator stresses for all asphalt mixtures tested was 0.988. Figure 6 shows the predicted and measured lateral deformation of a triaxial test specimen at failure. Fig. 7 gives the state of stress of a triaxial test specimen at failure as predicted by the finite element analysis.

Behavior of Specimens in Marshall Tests

The first problem encountered in the simulation of Marshall test was to choose the "correct" elastic modulus E. As E is a function of

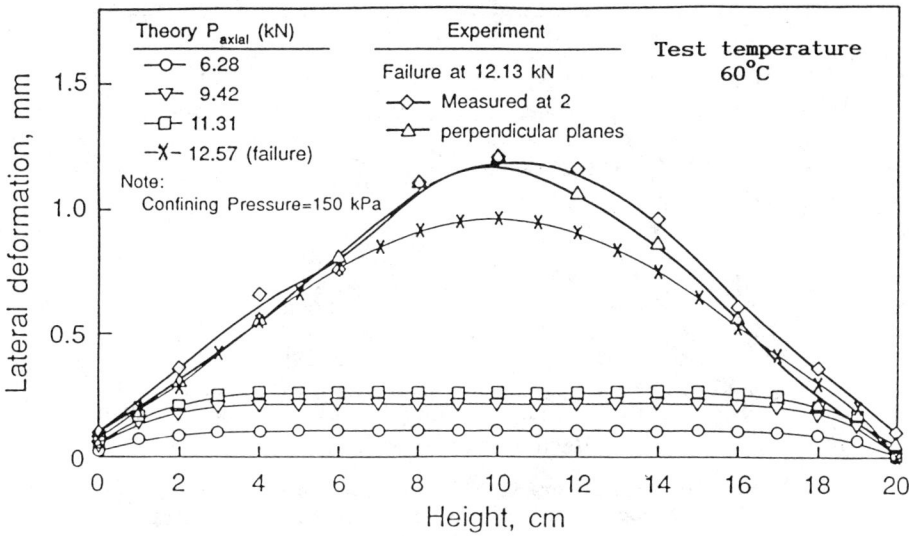

Fig. 6—Predicted and measured lateral deformation of asphalt specimen in triaxial test

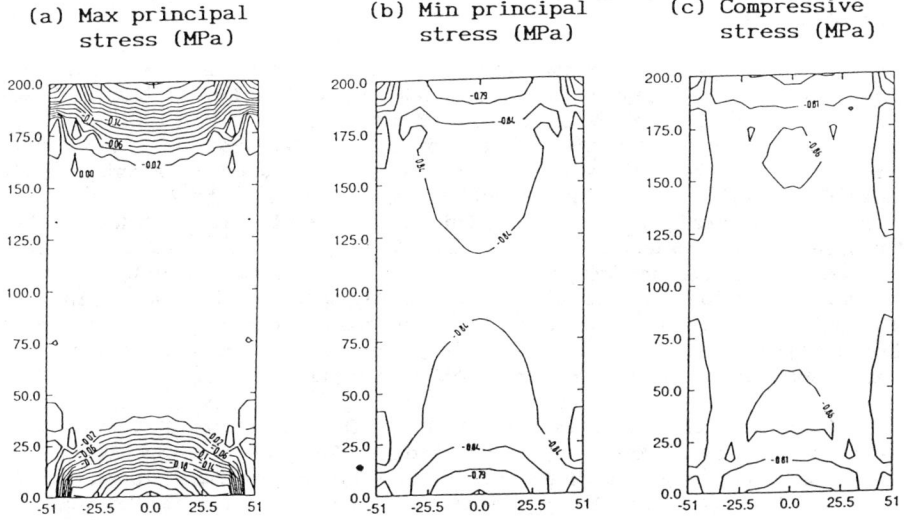

Fig. 7—Predicted stresses of triaxial test specimen at failure for the case of zero confining pressure

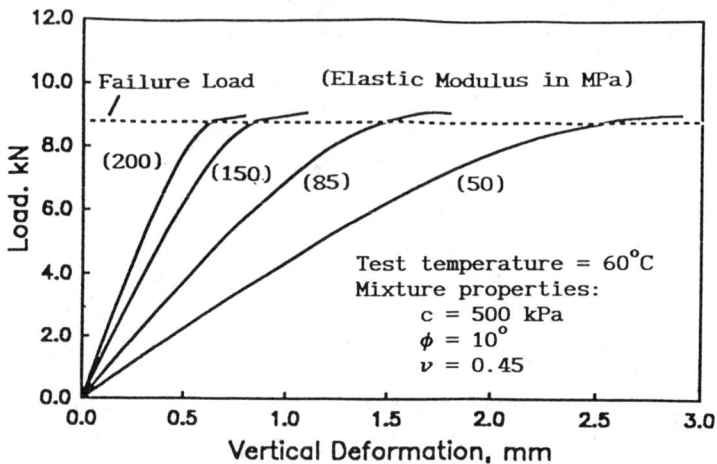

Fig. 8—Finite element predictions of Marshall stability using elastic moduli evaluated at different confining pressures

confining pressure, the problem becomes one to identify the confining pressure level at which the corresponding E would best simulate Marshall test behavior. As it turned out, as shown in Fig. 8, for a fixed value of c and ϕ, the Marshall stability value predicted by the finite element analysis remained unchanged regardless of the choice of E. This interesting finding highlights the value of mechanistic analysis. The computed stress contours within a specimen at failure are shown in Fig. 9.

Fig. 8 clearly shows that while the choice of elastic modulus has no effect on the computed value of Marshall stability, it has a direct effect on the magnitude of predicted flow value. Fig. 10 shows an example that compares the experimental load-deformation curves with those obtained by finite element analysis. It is of interest to note that Geotz [12] in a study investigating the use of triaxial test for asphalt mix design, had concluded from experimental evidence that triaxial tests conducted at 70 kPa (10 psi) confining pressure produced the same optimum binder content as that obtained by Marshall tests. An inspection of the test results such as that shown in Fig. 10 indicates that the use of 70 kPa confining pressure as suggested by Geotz appears to produce reasonable load-deformation curves for most cases when compared with experimental Marshall test results.

Based on the test results of all four mix types tested, the statistical coefficient of correlation r between the predicted and experimentally measured Marshall stability values was found to be 0.772. The corresponding r value for Marshall flow at failure was only 0.486. The finite element solutions typically under-estimate Marshall flow by about 20 to 35%. This is believed to be caused by the idealization in the Drucker-Prager model of elasto-plastic stress-strain behavior. The model does not represent correctly the

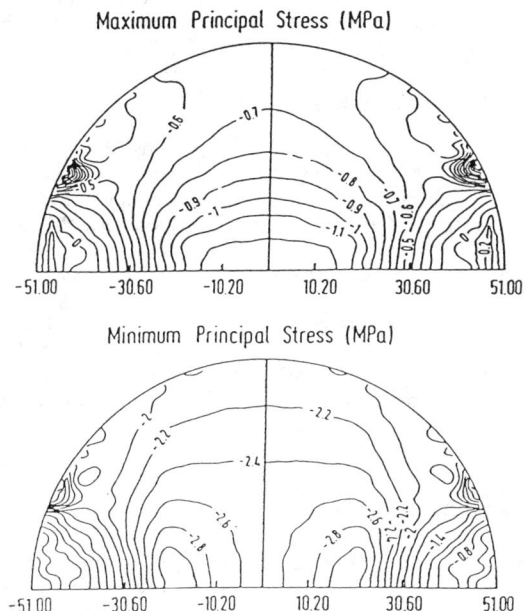

Fig. 9—Finite element prediction of stress contours in Marshall test specimen at failure

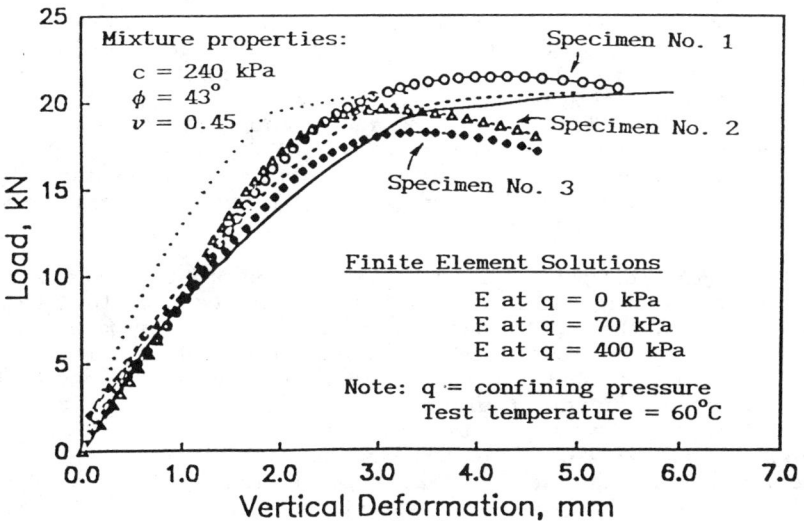

Fig. 10—Comparison of experimental load-deformation curves with finite element solutions

nonlinear stress-strain behavior in asphalt mixtures prior to reaching the peak load. This limitation, however, does not affect the application of the model structural design and analysis of pavements where linear elastic analysis is adopted.

Behavior of Specimens in Indirect Tension Tests

In the simulation of indirect tension tests, the elastic moduli obtained from triaxial tests with zero confining pressure were applied. Fig. 11 shows the results of measured and predicted behavior of an indirect tension test specimen. Good matching of stress-strain behavior was achieved until the peak load was reached. Reasonable agreement of vertical and horizontal deformation was also observed.

Of special interest to engineers performing the indirect tension test is the indirect tension strength of the specimen tested. The finite element analysis was able to provide very good estimates in this regard. The overall statistical coefficient correlation r between the predicted and measured indirect tension strength was 0.930.

SUMMARY AND CONCLUSIONS

This study has illustrated through simulations of stress-strain response of asphalt specimens in laboratory tests that triaxial test-determined properties can be employed to study the behavior of asphalt mixtures up to failure under three different loading conditions. The analyses were carried out using finite element models that incorporated the Drucker-Prager failure criterion. By matching with Mohr-Coulomb criterion, the Drucker-Prager criterion could be expressed in terms of triaxial test-determined properties c and ϕ.

Comparisons between predicted and measured laboratory test results show that very good estimates of the maximum deviator stress in triaxial tests and the indirect tension strength in indirect tension tests were obtained by the analyses based on triaxial test-determined properties. The corresponding coefficients of correlation for the two cases were 0.988 and 0.930. A slightly lower correlation coefficient of 0.772 was obtained for Marshall stability estimates, caused possibly by imperfect simulation of interface between specimen and test head.

Good matching of predicted and measured deformation was also observed in the initial stages of all tests. However, due to linear response assumed by the model adopted, larger discrepancies were found as the failure load was approached. In spite of this limitation, the proposed approach is still a useful tool for analyzing pavement responses under individual wheel loads where linear elastic models have been found to give acceptable results. It is noted that for application in analyzing pavement response under moving traffic loads, the described triaxial test could be supplemented by repeated-load triaxial tests to provide additional material parameters for this mode of loading.

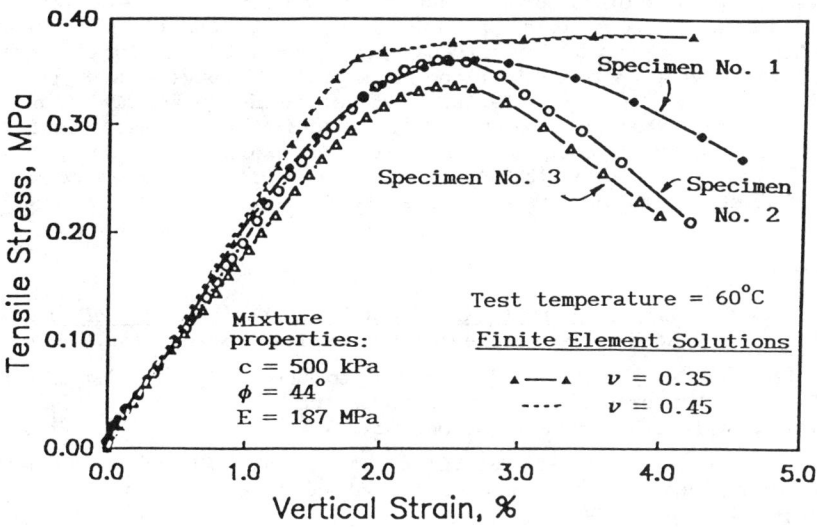

(a) Stress versus vertical strain

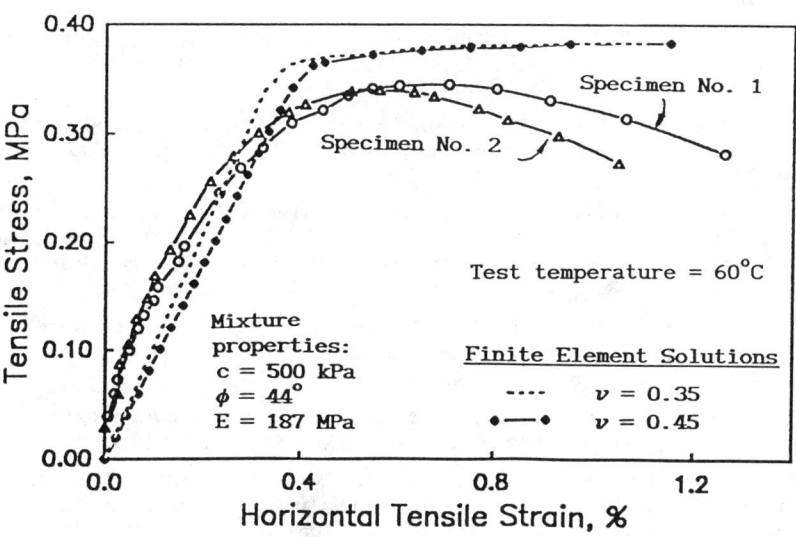

(b) Stress versus horizontal strain

Fig. 11—Predicted and measured stress-strain behavior of indirect tension test specimen

This study has provided positive evidence on the applicability of triaxial test-determined properties for examining the behavior of asphalt mixtures under loads in laboratory test conditions. These properties can be used for stress-strain behavior analysis in a mechanistic design and evaluation of structural pavement. It is significant that the same triaxial test-determined properties can be used to predict Marshall test and indirect tension test results, thereby providing a possible link between mechanistic design of asphalt mixes and the wealth of mix design experience associated with Marshall test accumulated over the past 50 years or so.

REFERENCE

[1] Nijboer L. W., Plasticity as a Factor in the System of Dense Bituminous Road Carpet, Elsevier Publishing Company, Inc., 1948.

[2] McLeod N. W., "A Rational Approach to the Design of Bituminous Paving Mixtures", Proc. Association of Asphalt Paving Technologists Vol. 19, 1950, pp. 82-187.

[3] Smith V. R., "Triaxial Stability Method for Flexible Pavement Design", Proc. Association of Asphalt Paving Technologists Vol. 18, 1949, pp. 63-94.

[4] Smith V. R., "Application of the Triaxial Test to Bituminous Mixtures", ASTM STP 106, American Society for Testing and Materials, 1951.

[5] Drucker D. C. and Prager W., "Soil Mechanics and Plasticity Analysis or Limit Design", Quarterly of Applied Mathematics, Vol. 10, 1952, pp. 157-165.

[6] Desai C. S. and Siriwardane H. J., Constitutive Laws for Engineering Materials with Emphasis on Geologic Materials, Prentice-Hall, Inc., Englewood Cliffs New Jersey, 1984.

[7] Cumaraswamy V., "Mechanical Behavior of Chemically Grouted Sand", J. of Geotechnical Engineering, Vol. 112, No. 9, 1986, pp. 869-887.

[8] Adachi T., Kimura M. and Tada S., "Analysis of the Preventive Mechanism of Landslide Stabilizing Piles", Third Int. Symposium on Numerical Models in Geomechanics, Elservier Applied Science, 1989, pp. 691-698.

[9] Fourie A. B. and Beer B. "An Illustration of the Importance of Soil Nonlinearity in Soil-Structure Interaction Problems", Computer and Geotechnics, Vol. 8, 1989, pp. 219-241.

[10] Asphalt Institute, "Mix Design Methods for Asphalt Concrete and other Hot-Mix Types", Manual Series MS-1, 5th Edition, 1983.

[11] Fwa T. F., Low B. H. and Tan S.A., "Compaction of Asphalt Mixtures for Laboratory Testing - Evaluation Based on Density Profile" Journal of Testing and Evaluation, Vol. 21, No. 5, September 1993, pp. 414-421.

[12] Geotz W. H., "Comparison of Triaxial and Marshall Test Results", Proc. Association of Asphalt Paving Technologists Vol. 20, 1951, pp. 200-245.

Richard L. Davis[1]

ENGINEERING PROPERTIES OF ASPHALT MIXTURES AND THEIR RELATIONSHIP TO PERFORMANCE

REFERENCE: Davis, R. L., "**Engineering Properties of Asphalt Mixtures and their Relationship to Performance,**" Properties of Asphalt Mixtures and the Relationship to their Performance, ASTM STP 1265, Gerald A. Huber and Dale S. Decker, Eds., American Society for Testing and Materials, Philadelphia, 1995.

ABSTRACT: Good performance for a highway, like any other product, is satisfaction of the user or customer. Most asphalt pavement users are well satisfied as long as the ride is smooth and the cost is low. The major purpose of this paper is to determine those engineering properties which result in a durable, smooth, low-cost asphalt pavement and to show how they can be controlled to assure continuing satisfaction of the road user. These properties are identified as being those that eliminate or reduce low-temperature cracking, reflection cracking, fatigue cracking, distortion and disintegration. Distortion or structural weakness can be eliminated by increasing the top size of the aggregate in the region of the weakness or increasing the thickness of the pavement depending on the type of distortion. Disintegration is controlled through softening the asphalt, increasing its content and the quality of the aggregate. Cracking is reduced through a combination of the foregoing factors. Proper management of these factors assures long term satisfaction.

KEYWORDS: engineering, science, mathematics, elasticity, viscoelasticity, cost, plasticity, mechanical properties, low-temperature fracture, reflection cracking, healing, fatigue, rutting, stripping, durability

Engineering is defined in the American Heritage Dictionary as "The application of scientific and mathematical principles to practical ends such as the design , construction, and operation of efficient and economical structures, equipment, and systems". Highway engineers have had great difficulty in satisfactorily defining the mechanical properties of asphalt mixtures because they consist of binder which varies from a soft viscous liquid at the highest pavement temperature to a brittle solid at the lowest pavement temperature mixed with broken stone of various gradations which is a hard solid at all road temperatures. For this reason the engineering properties of asphalt mixtures do not fall comfortably into the usual elastic design methods.While most engineers would prefer to

[1]Engineering Consultant, 1177 Pinewood Drive, Pittsburgh, PA 15243-1809

design an asphalt mixture in terms of linear elastic theory, there has been a realization that in hot weather conventional asphalt mixtures were being stressed beyond their elastic limit by modern truck traffic. This has resulted in attempts to use viscoelastic design methods. Viscoelastic design methods tend to further complicate mixture design which is already complicated by large sources of random variations in the most widely used test methods. If the elastic limit is increased to a level that not only removes the mixture from the critical area of frequent rutting problems but also gives it a comfortable cushion to take care of the variation due to random measurement errors, design using linear elastic methods would be valid and desireable.With the large reduction in rutting problems stability testing could be greatly reduced or even eliminated . But since there has been no acceptable method of measuring the elastic or proportional limit of asphalt mixtures and the whole concept of the elastic limit of asphalt mixtures has been shrouded in doubt, the tools have not been available to understand or to control it. The intent of this paper to both show how the elastic limit can be measured and how to increase it to comfortable levels.

The great quandary of science and mathematics has been that of bringing theory and experiment into harmony. The ancient Greeks recognized that all information which comes from the senses had error in it and therefore wanted to exclude all experimental data because they did not know how to cope with the error. It apparently was not as clear to them that the assumptions underlying all theory never are realized completely in practice. In other words, theory is a simplication of the complexity of the real world and so its application to the real world always involves error. Theorists often believe that theory is perfect and that any imperfection is in the physical world. Since the theorists tend to think that theory has little or no error and empiricism is invalid and empiricists tend to think that highly theoretical approaches are divorced from reality and the only fruitful course is experiment, it is little wonder that they grow farther and farther apart. What is the engineer to do to reach the practical ends called for in the definition of engineering given above? He must proceed to make the best use that he can of both theory and experiment while never losing sight of the fact that there is error and uncertainty in both which require proper understanding and management if the best use of the available materials is to be made.

A claim of knowledge in an area usually requires a suitable theory in order to be taken seriously, so even if the application of theory always has error in it, theory is still required. The fact that an engineer does not know anything until he has a theory does not protect him from choosing a theory that is more misleading than helpful. Many things that are known through faulty theory are not even approximately true and the application of theory is never without error and therefore uncertainty. There is no good method of estimating the size of the error due to theory and there should be. In the absence of a good method it is important to recognize that the application of theory involves error and an engineer must be constantly trying to evaluate the size and extent of this error. The hope is that recognition of error in both theory and practice will lead to a more realistic estimate of the sum of the errors through constantly checking theory against practice and practice against theory. An engineer should resist the tendency to overemphasize theory over experiment or experiment over theory.The essence of the scientific method is the realization that the optimum approach will employ a careful combination of the theoretical and the empirical.

The design of asphalt mixtures has been polarized by discussion of theoretical versus empirical approaches since its inception. Some highway engineers advocated a rational as opposed to a practical approach. Since even the most opinionated realized that it was difficult to support the argument that the practical approach did not involve reasoning, little has been said about rational designs in recent years. But the controversy continues with mechanistic versus statistical methods being advocated. The authors view is that it should first be recognized that most of our current problems of asphalt mixture design are due to trying to fit a material with an inadequate elastic limit to the requirements of modern traffic. Some designers have reduced the asphalt and increased the air voids in conventional asphalt mixtures in an attempt to raise the elastic limit. If this type of mixture is designed with too little asphalt in this manner, its durability is threatened by thin asphalt films, high air voids which increases the rate of oxidation, and high permability which makes it vulnerable to stripping. If designed with adequate asphalt and a proper level of air voids, the mixture is not stable enough to cope with modern truck traffic. No test method or level of testing can cure this problem. What is needed is an asphalt mixture with an elastic limit that is high enough to reduce the risk of rutting to an acceptable level while maintaining high asphalt film thicknesses and low air voids Such mixtures are possible and will be discussed later in this paper.

Conventional asphalt mixture design was marginal under the lower tire pressures of the past and has become inadequate for modern truck tire pressures. Measurement of properties can only be justified in critical areas of performance. An increase in the elastic limit of asphalt mixtures to a level that gives an adequate margin of safety or a low probability of failure could make much of our present testing unnecessary and allow the increase in asphalt content and the reduction of air voids that would greatly increase the durability of asphalt mixtures.

The idea of testing a small portion of a material and establishing its engineering or mechanical properties and then using mathematical relationships to design structures has long been used in the steel industry. Highway engineers have wanted to design highways in much the same manner, but it should be recognized that there are some marked differences between steel and asphalt mixtures. . Questions were raised as to the suitability of the theories of elasticity and of viscoelasticity for use with real materials when they were first proposed. Since both are derived from the partial differential equations of theoretical physics and rest on the assumption of continuous media, it was obvious that real materials were not really continuous. Actual practice has shown that theories such as the linear elastic theory which assumes homogeneity, isotropy, and continuity can be useful with some materials which do not strictly conform to the assumptions, but the uncertainty introduced by variation from the assumptions must be given careful consideration in an engineering analysis. The effect of this uncertainty will be addressed later in this paper.

HISTORICAL PERSPECTIVE

Man from the earliest times sought pathways that were firm and which could easily bear his weight. The better drainage of higher ground established the *high* ways of ancient times. With the advent of wheeled carts and carriages some areas in the traveled way were unable to support this new traffic and it was necessary to improve its ability to support the increased stress. Many of these roadways were only weak in rainy weather. Improved drainage was often sufficient to make them all-weather roads, but some required the addition of materials to the surface that were better able to resist deformation. This led to pavements which had as their function the ability to resist the high surface stress of wheeled vehicles and reduce the stress transmitted through the pavement to the point where the underlying natural earth could bear it without excessive deformation. This is still the essence of pavement design. These pavements were also designed to reduce the amount of moisture reaching the supporting earth beneath the pavement increasing its ability to support higher stresses. While granular materials could develop the necessary resistance to stress, asphalt mixtures were able to give not only increased protection to the soil beneath from stress but also from water.

The ability of HMA to resist permanent deformation at high temperatures has always been a concern. The early rock asphalt mixtures were composed of fine aggregate and it was noted early that the adjustment of asphalt content to precise levels was required if rutting was to be reduced. It was because of deformation problems with sheet asphalt pavements that F. J. Warren patented an asphalt pavement with stone in it in 1901. Warren felt that his patent covered all pavements in which any of the aggregate was retained on a 2.0E-3 m (10 mesh) sieve. The Warren Brothers Company brought suit in Federal Court in Topeka, Kansas against a number of parties in that area for infringing its patent. But after considering the high cost of such suits, George C. Warren decided to accept an agreement by the defendants to restrict their use of aggregates to those that would pass a half inch sieve. This was the origin of the term "Topeka Mix" from which evolved conventional HMA.

George C. Warren explained to the Board of Directors of the Warren Brothers Company that he really had not given up anything because a mixture with stone no larger than a half an inch could never support the heavy traffic for which the Warren pavement was designed. The reader should realize that heavy traffic of that day consisted of heavy wagons with narrow steel rimmed wheels and trucks with solid rubber tires which placed much higher stresses on the pavement than modern traffic does. Little did Mr. Warren foresee the balloon or pneumatic truck tire which was developed in the late 1920's and which made the use of HMA feasible under truck traffic.

One of the stated purposes of the balloon tire was to make it possible for a truck to drive over ordinary asphalt pavement in the hottest weather without noticeable deformation. After some research it was decided that 4.1E5 Pa (60 psi) was the highest pressure that could be tolerated at the highest summer temperatures. We see, some forty years, later at the time of the AASHO Test Road a tire pressure of 4.8E5 Pa (70 psi) was adopted. Had pressures of 1.0E6 Pa (150 psi) been used the results of the test road for asphalt pavement would have been completely different.

In the middle 1970's pavements began fail in a new way. This is in the period in which people began to speak of the "goodies" being removed from the asphalt. Knowing enough about the manufacture of asphalt to know that the "goodies" had never been there in the first place, the author sought other explanations. There were those who said these failures were due to "stripping", but this was hard to believe since they most often occurred during long, hot, dry periods. Also why should a pavement which had shown no problems with moisture for twenty years suddenly begin to fail in dry, hot weather. In searching for another explanation it was learned that many truckers had increased their tire pressures to the 8.6E5 to 1.0E6 Pa. (125 to 150 psi) range. Knowing of the results of the tests in the 1920's, it was immediately recognized that tire pressures had advanced beyond the ability of conventional HMA to resist such high stresses at the higher temperatures and presented a most serious problem unless tire pressures could be lowered. It is the author's belief that the rutting problem would disappear if tire pressures were lowered to 4.1E5 Pa (60 psi) today.

A tire pressure of 4.1E5 Pa (60 psi) was selected to make conventional asphalt mixtures structurally adequate at high summer temperatues. Highway engineers were alert to the effect of increased loads on pavements, but in the long period between the selection of this level of pressure and the 1970's during which truck tire pressures were held near the desired level the critical significance of tire pressure was either forgottened or never appreciated. So state highway departments were caught off guard and the increase in tire pressure was allowed with little or no protest while increase in load limits were being valiantly fought.

After determining that there was little chance of lowering tire pressures, It was recognized that the rutting problem was not only a great challenge but also a great opportunity. The easy solution was to put a small amount of something in the mix that would increase its elastic limit to the point that the higher tire pressures could be borne without permanent deformation. A search for such a material was started and a formulation was found that was technologically successful in coping with very heavy truck traffic even in the hottest weather but a cost analysis gave convincing evidence that it was too expensive to be commercially competitive. The definition of engineering given at the first of this paper states that engineering structures should be economical and engineering properties includes low cost. It is not meant to imply that this approach cannot be economically feasible, but merely that after considerable effort no economically competitive material was found in this study.

The teaching of history or more properly experience is that those asphalt mixtures that had elastic proportional limits or yield strengths that were high enough to keep the deformations due to highway traffic stresses within the linear elastic range were the most successful pavements. It is one of the major purposes of this paper to show how this can be accomplished at reasonable cost under our present day truck traffic.

THEORETICAL CONCEPTS

Newton developed a theory of gravitation to explain force at a distance between earth and celestial bodies. This led to potential field theory which Boussinesq [1] used in determining the distribution of stress in a linear elastic body. The usual assumptions of a linear elastic, homogeneous, isotropic semi-infinite half space simplified the analysis as did the application of a point load. A point load is a finite load that because a point has no area applies an infinite stress to the half space as shown in Figure 1.

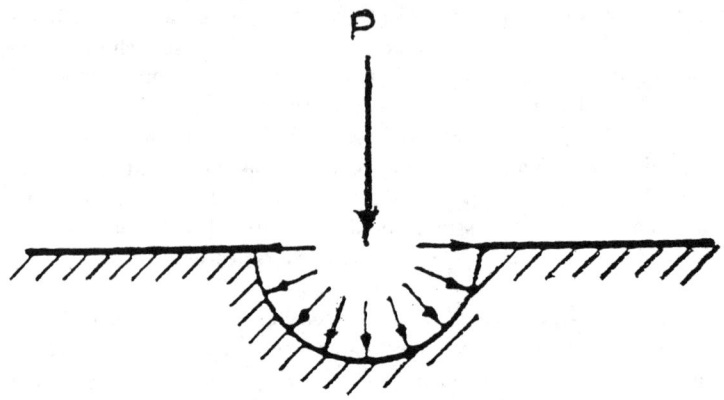

FIG.1 - *Application of a point load to a solid*

An infinite stress will plasticize any real solid since it would be far above its elastic limit. Since the half space is linear elastic, homogeneous, isotropic, continuous and semi-infinite, the plasticized volume would spread the stresses from the point of loading of the solid lowering the stress level until it reaches the elastic or proportional limit of the solid as shown in Figure 2.

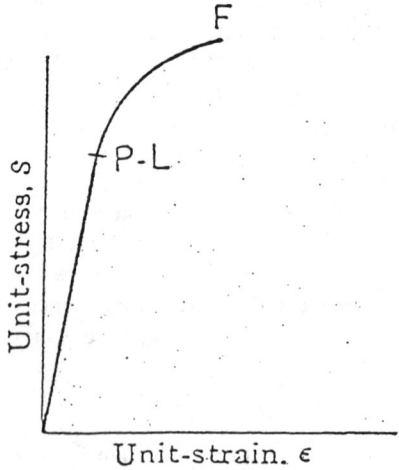

FIG. 2 - *Proportional limit of a solid*

The volume plasticized is determined by the elastic or proportional limit and the load. As the plasticized volume increases the surface area at the boundary between the plasticized material and the unplasticized material increases reducing the stress. When that stress falls below the elastic limit of the material, the rest of it resists elastically. The boundary conditions are that the stress on the surface of the half space is everywhere zero except at the point of loading where it is infinite and the stress is lowered as the distance from the point source increases within the half space to the point where it is zero at infinity.

Studies by the author have shown that the Boussinesq distribution of stresses for a linear elastic material are conservative for asphalt mixtures, the Boussinesq distribution is helpful in visualizing the mechanism of stress reduction with depth in an asphalt pavement..It should be recognized that good quality road stone will have compressive strength of 6.9E7 to 2.8E8 Pa (10,000 to 40,000 psi) which is far above the stresses applied by truck tires so it would not be plasticized. Asphalt has an elastic limit of about 25 Pa (3.6E-3 psi). Large size stone at the surface of the pavement can resist the high surface stresses and spread them to a larger area reducing the stress and increasing the elastic limit of the pavement in proportion to the increase in the size of the stone. The determination of elastic moduli for asphalt mixtures is greatly complicated by this great difference. The effect of specimen size and shape on the strength characteristics of asphalt mixtures is examined in the section on Engineering Properties.

This concept does not strictly apply to asphalt mixtures, yet it can illuminate how solids resist stresses and give an idea of how stresses are distributed in asphalt mixtures. An examination of Figures 1, 2, 3, and 4 shows how stress is distributed in an ideal elastic solid.

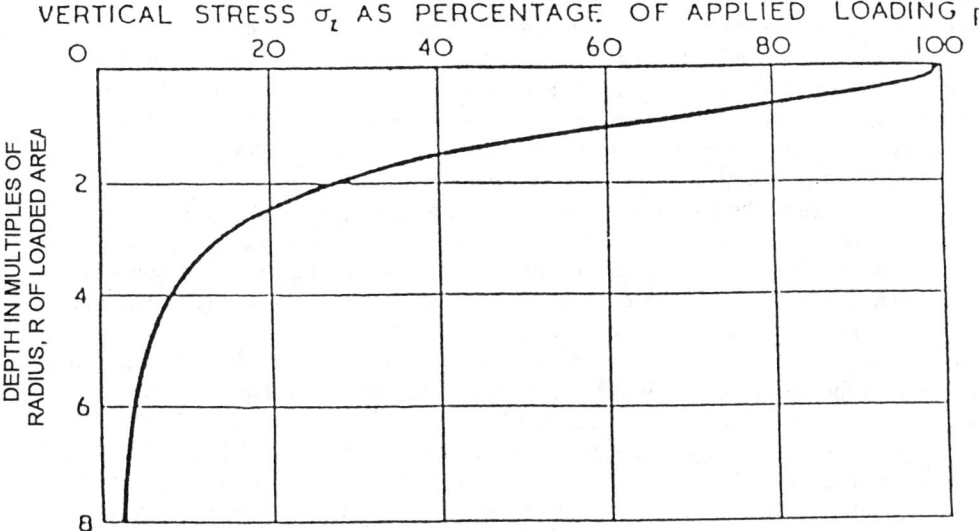

FIG. 3 - *Verticle stress on the axis within a semi-infinite elastic medium due to a circular uniform stress*

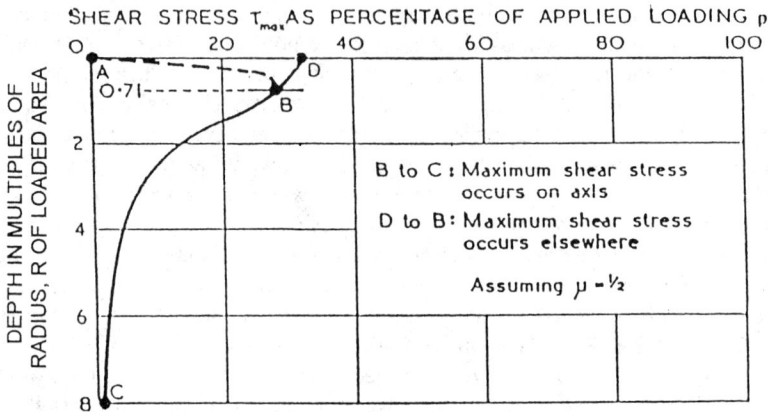

FIG. 4 - *Shear stress within a semi-infinite elastic medium due to a circular uniform stress*

Figure 1 shows what happens when the proportional limit in Figure 2 is surpassed as described above. This is what happens in asphalt pavements when the linear elastic range is exceeded. They are plasticized and flow. Figures 3 and 4 show how vertical and shear stresses are reduced with depth in multiples of radius of a uniform circular load in a linear elastic solid.

During the period between the advent of the balloon tire and the increase in truck tire pressures in the 1970's highway engineers became comfortable with the idea that load level controlled the structural requirements for pavement. This was while tire pressures were controlled in the region of 4.1 E5 Pa (60 psi). This was essentially true as long as the elastic limit of HMA was not exceeded by too large an amount. Figures 1 and 2 show that it is stress above the elastic limit that plasticizes a solid, not load. Loads no matter how great do not plasticize a solid as long as the stress is below the elastic limit. Load does have an effect on pavement structure. The ordinate of Figures 3 and 4 shows depth in multiples of the radius of the circular loading area. This area and therefore the radius is increased with increasing load if stress is held constant at the surface. Stress is increased at every depth in the pavement with increasing radius of the surface loading even though the surface stress remains constant since the stress reduction is in terms of radius of the circular area covered by the load. A pavement may have to be thickened or strengthened to cope with the higher stress deeper in the pavement with increased load even though the surface stess may not exceed the elastic limit of the asphalt mixture. This will be considered in more depth in the section on Engineering Properties.

For many engineers the essense of the definition of the engineering properties of asphalt mixtures is the establishment of the relationship between stresses and strains throughout the mixture. This is greatly complicated by the difference in properties between the asphalt binder and the stone particles. While asphalt mixtures cannot be said to be isotropic and homogeneous, elastic theory can be helpful as long as the proportional limit of the mixture is not exceeded.first looking at a less complicated idealized model will be followed here. An ideal, elastic solid will first be examined.

No real material is isotropic and homgeneous since all materials are made up of atoms and atomic structure is not isotropic or homogeneous in a strict sense. The amount of uncertainty in the application of elastic theory depends on how well the basic assumptions are met. This uncertainty is usually covered by a factor of safety or confidence interval. Some call this the "ignorance interval". Yet elastic theory can be helpful in understanding the structural properties of all solids including asphalt mixtures even though its use introduces uncertainty which should be evaluated in an analysis. The departure from the linear elastic models increases as the size of the stone becomes large in relation to the size of the critical stress area. For example, stone large in size in relation to the high stresses at the surface of an asphalt pavement can greatly increase its elastic limit. The concepts of point load and elastic half space can be helpful in visualizing what happens in stress and strain fields when a load is applied to a solid whether it is linear elastic or not.

In 1943 Burmister authored a paper [2] on a method for calculating the stresses and strains in a two layered system through the use of the mathematical theory of elasticity. Burmister carefully pointed out that his method applied only to ideal (linearly elastic) materials and that the engineer must judge the validity and application of the method to a material on the basis of actual performance and experience. Since the Burmister approach to layered systems is based on the assumption of linear elasticity, it is helpful in understanding the general concept of layered pavements but it suffers from the same problems as the Boussinesq approach to the distribution of stress.

PRACTICAL EXPERIENCE

The authors first experience in the design of asphalt mixtures left a lasting impression. It was a very hot summer day the asphalt and the sieved sizes of the aggregate were arranged in beakers on the laboratory table. A thumb pushed into the asphalt showed it to be soft and easily displaced, but the aggregate was hard and resistant to penetration. This has caused the author to always think first of changes in the aggregate structure when trying to increase resistance to deformation of asphalt mixtures at high temperatures. Since the resistance to plastic deformation is rarely a problem at low temperatures, only high temperature resistance to plastic deformation is relevant. On the other hand when increased resistance to water, weathering, and low temperature cracking are required, the authors thoughts turn to changes in the asphalt.

Being naturally inquisitive, the author has made it a practice since the earliest field experience to probe into pavements with a knife or an ice pick, and later with a sharpened screwdriver. This practice gives a fundamental appreciation of a pavement's ability to resist deformation. It quickly becomes apparent that an asphalt pavement is very hard in cold weather and can be quite soft in hot weather. Sand mixtures are easily penetrated at high temperatures but when a large piece of aggregate is reached the resistance is increased dramatically. It was such probing that first gave the author an appreciation of the increase in resistance to deformation that large pieces of aggregate can give to an asphalt pavement.

Another early experience that helped the author to better understand how the resistance of asphalt pavements to deformation could be increased happened in Western

North Carolina. A prominent politician returning from a meeting one rainy night was killed in an automobile accident. The accident occurred on a sharp mountain curve. There were a number of possible causes for the accident, but a newspaper reporter noticed that the pavement was black and shiny in the outer wheel path near where the car had left the road. The newspapers started a crusade to eliminate all shiny spots on asphalt pavements. The section of road where the accident occured was immediately sealed with a single seal coat and the shiny spot was eliminated. The next summer the outer wheel path began to get shiny again and the newspapers turned up the heat on the district highway engineer. There was much discussion of what was happening at this particular spot in the highway department. There was also concern because the newspaper and the general public were pointing out that there were many such shiny spots in the outer wheel paths on sharp mountain curves. Some of the engineers thought that the binder was not able to hold the aggregate under the scrubbing action of the solid axles of the trucks on sharp mountain curves.

The author placed a square foot patch in the wheel path during hot weather on one of the sharpest curves and both observed it closely and photographed it frequently. With such a small patch each piece of stone could be identified and observed individually. The nature of the problem soon became apparent. The stone was not being scrubbed off the pavement. It was being forced down into the pavement and the asphalt was being squeezed to the surface. It was evident that on the sharp mountain curves the load was shifted to the outer wheels of the trucks by centrifugal force causing higher tire pressures in the outer tires and increased pressures on the outer wheel path. The increased stress was more than the pavement could bear and the small stone was forced down into the pavement. After the nature of the problem became evident, a number of small patches were placed in the same wheel path. These patches consisted of various gradations, and various frictionally different aggregates with three types of binder. Within a few weeks of hot weather it was evident that those patches with larger stone were best able to resist the stresses imposed by the truck tires. The usual seal stone was closely graded in an effort to make it as near one-size as was practicable with 100% passing the $1.27E-2$ m (1/2 inch) sieve and with very little passing the number $4.75E-3$ m (No.4) sieve. It was nominally a $9.5E-3$ m (3/8 in.) seal stone. A closely graded aggregate was selected with 100% passing the $2.5E-2$ m (1 in.) sieve and little passing the $1.25E-2$ m (1/2 in.) sieve, nominally a $1.9E-2$ m (3/4 in.) stone. This was the only change in the design and it solved the problem. The first section was a straight seal covering the accident site. After a year of traffic including two hot summers with no re-occurrance of the shiny sections, several hundred miles of mountain roads were treated eliminating the shiny sections. Doubling the size of the aggregate had cured a persistent problem making a strong impression on the author and alerting him to the importance of stone size in resisting high stress in pavements.

Curiousity demanded an explanation that would explain how doubling the size of the aggregate enabled the pavement to resist the stress applied by truck tires. It was evident that the surface of the pavement was completely covered no matter what size stone was used. So while it would take four times as many particles of aggregate to completely cover the pavement as it would for particles twice as large, the entire surface would still be covered. What was the source of the increased resistance? After a little thought it was clear that while the exact same surface would be covered, the volume of the aggregate layer would be twice as great. This meant that twice as much work both plastic and elastic would have to be done to press the larger particles into the surface as for the smaller particles. This increased work both increased the elastic limit of the old pavement and increased the resistance of the newly applied layer of aggregate. Over the years the author has found that the increase in the elastic limit of a pavement is proportional to the increase in the size of the aggregate.

A few years later another experience involving the size of the aggregate reinforced the seal stone experience. Along the North Carolina Coast stretch many miles of beautiful beaches consisting of wind blown sand which have been pounded by the Atlantic Ocean for thousands of years. The sand has been polished and rounded by the elements to the point that it is practically spherical. Almost all of it will pass a 6.0E-4 m (30 mesh) sieve and be retained on a 4.3E-4 m (40 mesh) sieve. When dry it does not support automobile traffic very well so in order to open up the beaches for summer vacationists, it was necessary to open roads. While there was miles and miles of sand, the nearest deposits of aggregate of a size greater than sand were over 1.6E 5 m (100 miles) away. It was found that asphalt emulsion mixed with the beach sand enabled moving traffic to get over the beach roads very well. But these roads had such low bearing capacity that the wheels of a parked automobile would penetrate completely through the six inches of pavement in two or three hours in hot weather. For this reason no parking signs were posted every half mile along the pavement.

A parking lot was required at an information center on one of the beaches. The author remembering his experience in the mountains recommended that the usual beach sand mix be put down and a layer of large stone be rolled into this pavement on a hot day. Seventy five pounds per square yard of stone 100% passing the 4.5E-2 m (1 1/2 in.) sieve and retained on a 2.5E-2 m (1 in.) was rolled into the beach sand mix in the hottest part of the day. This combination successfully served as parking lot for years. Even the higher pressure truck tires did not indent the surface of the information center pavement.

The success in solving deformation problems with larger stone in the mix led the author to suggest that this be tried to reduce cracking in asphalt overlays over cement concrete pavements. The movement of cement concrete slabs causes cracking in asphalt mixture overlays within a short time after paving. The thought was that large stone would increase the length of the crack because it would have to go around each individual piece of stone. Trial sections were put down using a number of different binders, different types of aggregate, and different size aggregate. The cracks were mapped in the cement concrete pavement, both before the overlay was placed and a year after it was placed. The reduction in cracks coming through the overlay was greatest for the largest size stone and was more closely related to the size of the stone than any other factor.

ENGINEERING PROPERTIES

After reviewing both theoretical and practical approaches to pavement analysis, we must address the question of what is meant by the "Engineering Properties of Asphalt Mixtures". The author believes these are the properties that make it possible for asphalt mixtures to produce a road surface that is smooth and will remain smooth for many years under the expected traffic at the lowest possible cost including maintenance. The original smoothness is the result of the skillful placing of the asphalt mixture, but what are the factors that cause the loss of smoothness. These are usually such things as cracking, distortion, and disintegration. Cracking is frequently separated into that due to low temperature, that due to cracks in the subgrade or a previously laid pavement reflecting upwards, and that due to fatigue. Distortion is usually separated into that due to structural weakness in the underlying layers of the pavement and that due to the inability of the surface mixture to withstand the stresses applied at the surface. Disintegration is usually attributed to low asphalt content, hardening of the asphalt film through oxidation, stripping of the asphalt film due to water, and in time from the accumulation of all the effects mentioned above.

The important engineering properties of asphalt mixtures are those properties that insure the continued smoothness of the pavement through resisting the distress effects which are listed above. A more detailed examination of each of the sources of distress will be made to clarify what are the desireable engineering properties for asphalt mixtures.

The author was faced with controlling low temperature cracking of asphalt mixtures nearly thirty years ago in connection with the formulation of asphalt paving grades. A number of crude sources were available some of which were highly temperature susceptible. This made it difficult to avoid critically high stiffness and resultant cracking at low temperatures with asphalt meeting the normal specifications. While a number of factors contribute to the problem, a study was made that showed low temperature cracking could be controlled by holding the stiffness in the asphalt binder to less than 2.1E9 Pa s during the life of the pavement. A number of instruments were developed for measuring stiffnesses at low temperatures. After an understanding of the fundamentals of this type test method was reached, it was realized that the standard bituminous penetrometer could be easily adapted to make an excellent instrument for measuring low temperature stiffness. This adaptation allowed the low temperature stiffness to be run in all our asphalt laboratories on a routine basis. No low temperature cracking was encountered in twenty five years using this method [3] of controlling stiffness of asphalt while others who attempted to use these crude sources did experience severe cracking problems.

Cracks starting in the subgrade or old pavement and reflecting upward through a new asphalt pavement are difficult to counteract. The best way to deal with them is to treat the subgrade so the cracks will not form or in the case of an overlay over concrete, break up the old pavement into pieces whose longest dimension is less than 4.6E-1 m (18 in.). In the authors experience asphalt mixtures with large stone and soft asphalt present the greatest resistance to the reflection of cracks.

Fatigue is a complex matter even with metals, but when consideration is given to the fact that asphalt stiffness can vary by a factor of 10 million between the lowest temperature and the highest temperature that a pavement reaches, an entirely different domain is entered. Laboratory tests on asphalt mixtures run at lower temperatures can certainly measure fatigue properties under those conditions, but they have little application at the highest temperatures to which asphalt pavements are subjected. This may explain why early predictions of fatigue life based on laboratory tests at temperatures far below the highest reached in the pavement were low by a factor of 700 to 1000. Field experience has caused the author to emphasize the importance of healing in delaying the appearance of cracking in asphalt pavements. During cold weather traffic stresses and strains may cause microcracks to form in asphalt pavements, this causes a reduction in density in the asphalt pavement. These microcracks will remain as long as the asphalt is hard just as they do in metals.When hot weather comes the asphalt binder is softened and the traffic stresses and strains recompact the pavement much as the roller did when the pavement was first placed. Metals subjected to the same temperature range as asphalt pavements do not generally soften in this manner. The properties of fatigue displayed by metals do not apply to asphalt mixtures which are recompacted and healed by higher summer temperatures and traffic stresses. Softer asphalt aids this process as does increased stress or pressure from traffic. The bottom of a very thick asphalt pavement does not experience temperatures as high those at the surface nor are the compactive stresses as high as those in the top part of the pavement. Therefore, conditions are less favorable for healing and microcracks may not be healed as completely at the bottom of the pavement as they are in the top portion of the pavement. However, softer asphalt not only tends to heal cracks at the surface of the pavement but also to reduce the incidence of cracks anywhere in the pavement. The presence of large size stone is also a barrier to the formation of cracks as noted earlier.

Distortion of the surface of an asphalt pavement can be due to weakness anywhere in the pavement structure. Fundamental to pavement design is the recognition not only that the surface must bear the stresses applied by traffic but also that every point beneath the surface of the pavement must be able to bear the stress applied to it. Figures 3 and 4 show that both vertical and shear stresses are rapidly reduced with increasing depth. The highest vertical stress is at the surface of the pavement and the highest shear stress is at the surface of the pavement at the edge of the loaded area. This means that the material at the surface must be highly resistant to deformation stresses but with increase in depth material of lower resistance will be adequate. This allows lower strength and lower cost materials to be successfully used down in the pavement. However, the materials used must be able to cope with the stress applied to them without excessive permanent deformation. The rutting problem can be divided into that due to the surface layer being unable to withstand the stresses applied by traffic which leads to plastification of the top $7.6E-2$ to $1.0E-1$ m (3 to 4 in.) and that due to permanent deformation because the subgrade or one of the lower layers is unable to support the stress. The first condition is called rutting of the first kind and the second is called rutting of the second kind in the rest of this paper.

The measurement of yield strength, elastic limit, proportional limit and bearing capacity of asphalt mixtures, where all of these terms are taken as having essentially the same meaning, has been regarded as difficult to impossible by many researchers. Yet it is important not only in the understanding of the rheology of asphalt mixtures but also of asphalt itself as well as polymer blends with asphalt. With a little patience it can be measured for all of these materials. All of these terms are measures of the stress level at which linear elastic recovery ends and permanent deformation begins in a material. Because this stress can be very low in some of the materials named above there is a popular belief that it cannot be measured. This is not true but since this paper is about asphalt mixtures the discussion will be limited only to this material. The author's efforts to measure the fundamental properties of asphalt mixtures started by testing 0.1×0.2 m (4 x 8 in.) specimens in triaxial compression but stopped for two reasons. One was due to difficulty in determining realistic lateral pressures and the other was difficulty in finding any relationship between triaxial test results and the deformation resistance of actual pavements.

After considerable research with various thicknesses of unconfined specimens an understanding of fundamental relationships of stress, strain and time began to evolve. Figure 5 shows the relationship between strain and time when a constant compressive stress is applied to a cylinder of asphalt mix. The slope of the line shows the viscosity of the mix at any time. It is evident that the viscosity of the mix increases with time and strain. The work done on the mixture hardens and compacts it raising its elastic limit. As the elastic limit increases a larger and larger part of the stress is supported elastically causing the line to flatten and the viscosity to increase. The stress continues to do work on the mixture raising its viscosity and its elastic limit and the line continues to flatten until the mixture reaches a level of elastic strength that can support the applied stress in a linear elastic manner. At this point there is no increase in strain with time, the line is horizontal and the viscosity is infinite. The stress continues to do work on the mixture raising its viscosity and its elastic limit and the line continues to flatten until the mixture reaches a level of elastic strength that can support the applied stress in a linear elastic manner. At this point there is no increase in strain with time, the line is horizontal and the viscosity is infinite.

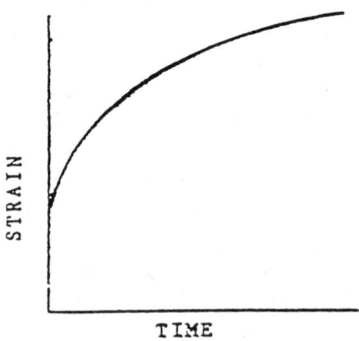

FIG. 5 - *Deformation of asphalt mixture at stress below elastic limit.*

If a higher level of stress is applied, work hardening will occur again and both the viscosity and the elastic limit for the mixture will increase. However, there is a limit to the elastic strength of the mixture as there is for all materials if this limit is exceeded the condition shown in Figure 6 occurs.

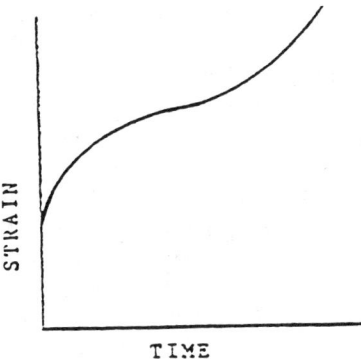

FIG. 6 - *Deformation of asphalt mixture at stress above the elastic limit.*

The stress is greater than the ultimate elastic limit of the mixture, the mixture is work hardened to its ultimate elastic strength, the line passes through the horizontal and the slope increases, the voids increase, the mixture is weakened. and fails rapidly. The Poisson's ratio will increase as the line flattens to the horizontal where it is equal to 1/2. If the stress is greater than the ultimate elastic limit and the line turns upward as in Figure 6, Poisson's ratio reaches values higher than 1/2.

After developing the necessary background for the measurement of bearing capacity, elastic limit or proportional limit, the effect of specimen dimensions on this property was investigated. It was found that shear strength and bearing capacities are highly dependent on the dimensions of the specimen tested. This is not surprising but it is important to include this knowledge in any attempt to understand the relationship between test results from a specimen and the ability of an actual pavement to resist deformation. For instance, it was found that a 0.1 x 0.2 m (4 x 8 in.) specimens when tested in axial unconfined compression at 60 C (140F) had bearing capacities of 2E4 to 7E4 Pa (3 to 10 psi) because the specimens dimensions did not provide sufficient confinement to work harden the asphalt mixture. When the dimensions of the specimens were changed to 0.1 x 0.1 m (4 x 4 in.) with the same mix the bearing capacity increased to about 5E4 to 12E4 Pa (7 to 15 psi). .On decreasing the thickness of specimens to dimensions of 0.1 x 0.05 m (4 x 2 in.) with no other change the bearing capacity increased to 1.4E5 x 2.8E5 Pa (20 x 40 psi). Changing the dimension of the specimen to 0.2 x 0.05 m (8 x 2 in.) resulted in bearing capacities of 1.0E6 to 1.4E6 Pa (150 to 200 psi). The aggregate in the asphalt mixture was densely graded with 100 % passing a 1.9E -2m (3/4 in.) sieve. It is evident that the elastic limit increases dramatically as the thickness of the specimen approaches the dimension of the largest aggregate and as confinement increases. .

Early work with triaxial compression specimens indicated that the effect of the size of the aggregate becomes negligible when it is less than one sixth the least dimension of the specimen. This is supported by the investigation in the preceding paragraph. But it should be remembered from Figures 3 and 4 that the highest stresses are at the surface of the pavement and they decrease rapidly with depth. This means that while a high elastic limit at the top of the pavement is necessary to cope with high traffic stresses it is not required deep in the pavement. Rutting problems of the first kind occur in the top 1.0E-1 to 1.3E-1 m (4 to 5 in.) because the stress reduces below the elastic limit of conventional HMA in that distance. A mixture with a higher elastic limit and modulus tends to spread the stress more rapidly than a weaker mixture so in general 5.1E-2 to 1.0E-1 m (2 to 4 in.) of new higher strength mixture is sufficient to alleviate most rutting problems of the first kind. The method outlined by E. S. Barber [4] to calculate equivalent thicknesses has been found useful even though any method based on the theory of elasticity has obvious limitations.

These principles are basic to understanding how asphalt mixtures can be work hardened or compacted to resist deformation and the limit that can be achieved by this process. This information is essential for understanding the design of asphalt mixtures capable of coping with high stress.

This forgoing creates real problems for any system of asphalt mixture design that attempts to measure the shear strength or any other strength property of asphalt mixtures containing aggregate with a standard size specimen and then apply these measurements to varying thickness of pavement. This is where asphalt mixture design becomes so different from steel design methods where the size of steel crystals seldom approach the thickness of steel members. Asphalt pavements are generally applied in layers of between 0.05 to 0.1 m. (2 to 4 in.) thickness. Much of the difficult task of relating the measured properties of specimens to asphalt pavements could easily be eliminated by testing specimens in the laboratory which have the actual thickness of the pavement..

The test method which the author found best is very simple. It requires two steel plates, one 3.0E-1 m (12 in.) diameter and 3.2E-2 m (1.25 in.) thick, the other 3.6E-1 m (14 in.) diameter and 2.5E-2 m (1 in.) thick. The 3.6E-1 m (14 in.) diameter plate has a plastic band around it to hold the asphalt mixture on the plate. Enough mix is evenly distributed over the plate to make a compacted specimen the thickness of the pavement.

The 3.0E-1 m (12 in.) diameter plate is centered over the loose mix and both plates and the mix are brought to a temperature equal to the highest pavement temperature expected in the design life of the pavement. This temperature is maintained throughout the test. A stress of 6.9E4 Pa (10 psi) is applied to the mix through the 3.0E-1 m (12 in.) plate. As the mix is compacted, the distance between the plates is recorded at 30 seconds after the stress is applied, and at intervals of one minute through ten minutes. Figure 5 shows a typical result. The temperature is maintained, the stress is increased and the procedure is repeated until a stress that results in a plot like that in Figure 6 is reached. No compaction is necessary prior to running the test. This is the procedure described earlier in connection with Figure 5 and 6. Compaction or work hardening is an integral part of the test procedure. The 3.0E-1 m (12 in.) plate approximates the area covered by a truck tire. The test result is the elastic limit of the asphalt mixture under conditions approximating those in the actual pavement. The elastic limit so measured should be twice the highest tire pressure expected on the pavement. The ability of the surface underneath the pavement tested above should be evaluated for its ability to support the top layer of pavement. For those who would like more information on this test method and some test data see reference [5].

In order to more nearly approach actual road conditions the author has made it a practice to further evaluate a selected pavement design in square sections 1.8 m (6 ft) on a side. These have usually been placed where trucks with controlled tire pressures at a quarry or hot plant go over them just before getting on the scale. In this way, it is easy to keep track of the number of passages and weight of the trucks travelling over the sections and to see how each pavement section responds to the traffic.

The best way to protect asphalt pavement from those factors that lead to disintegration is to increase the asphalt content and decrease the air voids in the compacted mixture. There has been a tendency on the part of some designers to reduce the asphalt content and increase air voids in an attempt to produce mixtures capable of resistiing traffic stresses without excessive deformation. This does increase both the viscosity and the elastic limit of the mixture and slow down the failure depicted in Figure 6. This action does not usually increase the elastic limit of conventional HMA to a level where it can halt continuing compaction under traffic stresses but it does flatten and extend the horizontal segment of the line and extend the time before disintegration occurs. If the elastic limit is increased to a high enough level, the voids in the mineral aggregate can be filled with asphalt increasing asphalt film thickness and dramatically reducing the effects of oxidation and stripping while still holding the elastic limit to a level where catastrophic failure is avoided. If the elastic structure of the mixture is improved sufficiently, softer asphalt can be used improving the mixture's ability to heal itself. These factors greatly increase the useful life of the pavement.

CONCLUSION

Engineers must make the optimum use of both theory and experiment in the design and application of asphalt mixtures which is especially difficult because of the essential nature of these mixtures. Increasing the elastic limit to a level twice the highest tire pressure expected on the pavement vastly simplifies the process.

REFERENCES

[1] Boussinesq, J., "Application des Potentiels a l'etude de l'equilibre et du Mouvement des Solides Elastique, Gauthier-Villars, Paris, 1885.
[2] Burmister, D. M., "The Theory of Stresses and Dispacements in Layered Systems and Applications to the Design of Airport Runways," Proceedings Highway Research Board, Vol. 23,1943.
[3] Davis, R. L., "Reduction of Low-Temperature Cracking in Asphalt Pavements" Proceedings of Canadian Technical Asphalt Association, Vol. 33, 1988.
[4] Barber, E. S., Proceedings Highway Research Board, Vol. 23, Page 146, 1943
[5] Davis, R. L., "Relationship Between the Rheological Properties of Asphalt and the Rheological Properties of Mixtures and Pavements," *Asphalt Rheology, Relationship to Mixture*, ASTM STP 941, O. E. Briscoe, Ed. American Society for Testing and Materials, Philadelphia, 1987, pp 28-50.

Shin-Che Huang,[1] Mang Tia,[2] and Byron E. Ruth[2]

EVALUATION OF AGING CHARACTERISTICS OF MODIFIED ASPHALT MIXTURES

REFERENCE: Huang, S.-C., Tia, M., and Ruth, B. E., **"Evaluation of Aging Characteristics of Modified Asphalt Mixtures,"** Engineering Properties of Asphalt Mixtures and the Relationship to their Performance, ASTM STP 1265, Gerald A. Huber and Dale S. Decker, Eds., American Society for Testing and Materials, Philadelphia, 1995.

ABSTRACT: A laboratory investigation using typical Florida dense-graded and open-graded mixtures was conducted to evaluate the effects of a few promising modifiers on the long-term aging characteristics of asphalt mixtures. Six types of modifiers including gilsonite, carbon black, fine ground tire rubber (GTR-80), styrene-butylene-rubber (SBR), ethylene vinyl acetate (EVA) and styrene-ethylene-butylene-styrene (SEBS) were used in this study. These modifiers were blended with AC-30, AC-20, and AC-5 base asphalts to produce a total of ten modified asphalts. These ten modified asphalt blends and the unmodified AC-30 were used to make Florida type S-1 structural course and type FC-2 friction course asphalt mixtures. These mixtures were compacted into Marshall specimens for age hardening and subsequent testing. A total of 416 compacted modified and unmodified asphalt mixture specimens were subjected to a forced-draft oven aging process for 90 days at 60 °C (140 °F), and natural sunlight for 6 months to simulate the long-term aging effects on these mixtures. These aged and unaged asphalt mixtures were evaluated by (1) resilient modulus tests at 5 and 25 °C (41 and 77 °F) and (2) indirect tensile tests at 5 and 25 °C (41 and 77 °F). The effects of cycle frequency used in the resilient modulus tests were also assessed. The test results indicate that the modified asphalt mixtures appear to have more pronounced delayed elastic behavior as compared with that of the unmodified asphalt mixtures. The fracture energy, which is characterized by means of the area of stress-strain curve, indicated that the modified binders using a base asphalt of AC-30 tend to be too brittle as they harden further with time. However, modified binders using a base asphalt of AC-5 appear to be softer than that of conventional AC-30 under typical Florida conditions.

KEYWORD: gilsonite, carbon black, fine ground tire rubber (GTR-80), coarse ground tire rubber (GTR-40), styrene-butylene-rubber (SBR), ethylene vinyl acetate (EVA), styrene-ethylene-butylene-styrene (SEBS), long-term aging

---

1   Postdoctoral associate of Civil Engineering Department, University of Florida, Gainesville, Florida
2   Professors of Civil Engineering Department, University of Florida, Gainesville, Florida

INTRODUCTION

In recent years, more and more asphalt modifiers have been used in asphalt paving mixtures in an attempt to improve the flow characteristics and temperature susceptibility of the asphalt binders, and subsequently to improve the performance of the asphalt pavements. Ground tire rubber (GTR) has been successfully incorporated in dense-graded friction course mixtures, open-graded friction course mixtures and in asphalt-rubber binder for stress-relief membrane interlayers in Florida. However, the long-term durability of these GTR-modified asphalt binders is still unknown and needs to be evaluated. Other asphalt modifiers which are of potential usage include (1) styrene butadiene rubber (SBR), (2) ethylene vinyl acetate (EVA), (3) styrene ethylene butylene styrene (SEBS), (4) gilsonite and (5) carbon black. While these modifiers could offer some improvement in short-term performance characteristics, the long-term durability of these modified binders is still not well understood.

The main objective of the research to be presented in this paper is to determine the aging characteristics of typical Florida dense-graded and open-graded paving mixtures incorporating these modifiers.

EXPERIMENTAL TEST PROGRAM

The asphalt modifiers used in the testing program include a gilsonite, a carbon black, a fine ground tire rubber passing #80 sieve (GTR-80), an elastomeric polymer styrene-butadiene rubber latex (SBR), a copolymer ethylene vinyl acetate (EVA) and a block copolymer styrene-ethylene-butylene-styrene (SEBS).

These six modifiers were blended with AC-30, AC-20 and AC-5 base asphalt cements to produce a total of ten modified asphalt blends. These ten modified asphalt blends and an unmodified AC-30 base asphalt were used to make Florida type S-1 structural course and type FC-2 friction course mixtures. Table 1 shows the aggregate gradations for the S-1 and the FC-2 mixes. These mixtures were compacted into Marshall specimens for age hardening and subsequent testing. The compacted modified and unmodified asphalt concrete mixtures were subjected to a forced-draft oven at 60 °C (140 °F) for 90 days and natural sunlight for 6 months. Some unaged specimens were also tested for comparison purpose. Results of previous asphalt aging studies [1] have indicated that exposure of Marshall specimens to forced-draft oven at 60 °C for 28 days is roughly equivalent to exposure of Marshall specimens to natural sunlight for 6 months or 12 to 24 months of field aging in a well compacted pavement. An exposure to a forced-draft oven for 90 days might be equivalent to 3-6 years of field aging. These aged and unaged asphalt mixtures were evaluated by (1) Resilient modulus tests (ASTM D-4123) at 5° and 25 °C, (2) Indirect tensile tests at 5 °C (41 °F), and (3) Indirect tensile tests at 25 °C (77 °F). Two replicate specimens were tested for each combination of test parameters. The total number of test specimens amounted to 416.

RESULTS AND DISCUSSION

EVALUATION OF RESILIENT MODULUS

The resilient modulus of an asphalt concrete can directly be related to the structural contribution of that layer or that component of the pavement. The resilient modulus is defined as the ratio of applied stress to corresponding recoverable strain during repeated loading. Mixtures with high resilient modulus at the low temperature are considered to be too brittle to resist cracking caused by thermal and

TABLE 1--*Job mix formula and aggregate types for both dense-graded and open-graded mixes*

	Dense-Graded Mix	Open-Graded Mix
Aggregate types	20% #67 Stone 25% S-1B 35% Screening 18% Sand 2% Mineral Filler	87% FC-2 Rock 11% Screening 2% Mineral Filler
Sieve size	% passing	
3/4" (19 mm)	100%	100%
1/2" (12.5 mm)	97.08%	100%
3/8" (9.5 mm)	88.98%	98.7%
#4 (4.75 mm)	60.64%	16.45%
#10 (2 mm)	45.24%	11.55%
#40 (0.425 mm)	25.33%	6.83%
#80 (0.18 mm)	8.58%	4.23%
#200 (0.075 mm)	4.24%	2.63%

load effects. It also has been found that the resilient modulus versus temperature relationship could be employed to aid the selection of the proper asphalt concrete mixtures to be used over the expected temperature range where the pavement will face in the field. In this test, a pulsating haversine load of approximately 15 percent of the failure load is applied to the specimen at a loading frequency of 0.33, 0.5, and 1 Hz with a load duration of 0.1 second (i.e., the corresponding resting times were 2.9, 1.9 and 0.9 second). The resulting horizontal deformation of the specimen is measured and, with an assumed Poisson's ratio, is used to calculate the resilient modulus.

Effects of Cycle frequency

Statistical analysis using the ANOVA technique was employed to determine the significance of the effects of test cycle frequency on the measured total resilient modulus. Results from ANOVA technique indicate that the effect of the test cycle frequency on the measured total resilient modulus at temperature of 5 °C (41 °F) is not significant for either the unmodified or the modified asphalt mixtures. It appears that the material has less delayed elastic response at this low temperature. However, at temperature of 25 °C (77 °F), the resilient modulus appears to increase slightly as the cycle frequency increases. This observation is more noticeable among the polymer modified asphalt mixtures, such as GTR-80 and SEBS modified asphalt mixtures. This can be explained by the fact that a higher test cycle frequency would result in a shorter resting time for the loaded specimen to rebound elastically between the pulsed loads. If a tested specimen does not rebound completely, the total recovered strain would be lower and the determined resilient modulus would be higher. Modified asphalts with more pronounced delayed elastic behavior would be affected more by the shortening of the resting time.

Comparison Between Dense-Graded and Open-Graded Mixtures

The results of resilient modulus tests on the unaged dense-graded and open-graded mixtures are displayed in Table 2. The average resilient modulus values from tests run at 0.33, 0.5 and 1 Hz were used.

All of the modified open-graded asphalt mixtures show lower value of resilient modulus than that of dense-graded asphalt mixtures at both 5 °C (41 °F) and 25 °C (77 °F). This behavior is reasonable and as anticipated. The higher air void content in the open-graded mix would cause a higher recovered strain, and therefore the determined resilient modulus would be lower.

The rate of age-hardening of asphalt mixtures caused by the forced-draft oven process are characterized by means of the ratio of resilient modulus after and before the aging process. Table 3 shows the ratio of resilient modulus for the dense- and open-graded mixtures at temperatures of 5 °C (41 °F) and 25 °C (77 °F). As seen from this Table, the rate of age-hardening of open-graded mixtures is lower than that of dense-graded mixtures. It appears that the aggregate gradation along with proper mix design play an important role on the pavement age-hardening. In addition, all of the modified open-graded asphalt mixtures show slower rate of age-hardening than that of conventional unmodified AC-30 asphalt mixture as seen from the ratio of resilient modulus at 5 °C (41 °F). On the basis of the ratio of resilient modulus at 25 °C (77 °F), all of the modified open-graded asphalt mixtures, with the exception of those containing EVA and SEBS modified AC-5, also show slower rate of age-hardening than that of the unmodified AC-30 asphalt mixture. It appears that the combination of proper modified asphalt binders and proper aggregate type and gradation could slow down the age-hardening of the asphalt pavement.

TABLE--2 Resilient modulus, indirect tensile strength, strain at failure, and fracture energy of unaged dense-graded and open-graded mixtures.

Asphalt Type	5°C				25°C			
	Total Resilient Modulus (ksi)	Indirect Tensile Strength (psi)	Strain at Failure (%)	Fracture Energy (Pa)	Total Resilient Modulus (ksi)	Indirect Tensile Strength (psi)	Strain at Failure (%)	Fracture Energy (Pa)
AC-30	1353* (923)**	473.8 (293.7)	0.045 (0.058)	1096 (884)	524 (434)	159.7 (123.0)	0.285 (0.155)	1919 (776)
+ Carbon Black	1568 (915)	474.0 (279.9)	0.055 (0.051)	1386 (763)	729 (476)	197.7 (141.0)	0.313 (0.181)	2470 (1028)
+ Gilsonite	1673 (1112)	483.1 (303.3)	0.054 (0.055)	1375 (869)	697 (564)	239.9 (160.4)	0.239 (0.121)	2346 (786)
+ GTR-80	1353 (1102)	441.0 (337.8)	0.069 (0.059)	1637 (1039)	461 (499)	160.3 (150.5)	0.338 (0.150)	2147 (869)
+ SBR	1240 (928)	425.9 (279.4)	0.081 (0.070)	1807 (1041)	623 (463)	164.7 (132.2)	0.263 (0.175)	1713 (920)
+ EVA	1812 (1112)	514.7 (334.3)	0.044 (0.057)	1189 (992)	803 (557)	247.5 (163.3)	0.209 (0.115)	2077 (752)
+ SEBS	1880 (1085)	490.5 (281.1)	0.046 (0.059)	1172 (828)	982 (587)	259.8 (168.5)	0.194 (0.106)	2064 (723)
AC-20+ EVA	1473 (804)	408.5 (312.1)	0.056 (0.069)	1517 (1056)	777 (438)	200.2 (118.7)	0.224 (0.144)	1731 (675)
AC-20+ SEBS	1607 (1037)	456.7 (274.7)	0.076 (0.057)	1794 (851)	780 (548)	213.7 (146.3)	0.235 (0.121)	2111 (683)
AC-5+ EVA	1298 (765)	454.5 (251.5)	0.128 (0.097)	3159 (1308)	601 (340)	151.3 (90.3)	0.212 (0.138)	1292 (493)
AC-5+ SEBS	1073 (881)	408.6 (251.6)	0.118 (0.059)	2507 (799)	444 (373)	141.9 (97.3)	0.250 (0.156)	1359 (617)

*   Values for Dense-graded mixes
** Values for Open-graded mixes

TABLE 3--Aging indexes of dense-graded and open-graded mixtures after a 90 day forced-draft oven aging

Asphalt Type	5°C (41°F)				25°C (77°F)			
	Resilient Modulus	Indirect Tensile Strength	Failure Strain	Fracture Energy	Resilient Modulus	Indirect Tensile Strength	Failure Strain	Fracture Energy
AC-30	1.23 (1.27)	0.91 (0.81)	0.73 (1.09)	0.67 (1.06)	1.91 (1.44)	1.64 (1.49)	0.65 (0.63)	1.04 (0.90)
+ Carbon Black	1.05 (1.20)	0.91 (1.06)	0.89 (0.88)	0.79 (0.92)	1.40 (1.28)	1.40 (1.39)	0.60 (0.63)	0.85 (0.87)
+ Gilsonite	1.10 (1.04)	0.91 (0.98)	0.51 (0.69)	0.48 (0.70)	1.74 (1.27)	1.41 (1.41)	0.72 (0.75)	1.00 (1.03)
+ GTR-80	1.18 (1.05)	0.94 (0.92)	0.84 (0.74)	0.75 (0.69)	1.77 (1.35)	1.53 (1.30)	0.59 (0.62)	0.95 (0.85)
+ SBR	1.41 (1.16)	1.13 (1.10)	0.61 (0.61)	0.69 (0.65)	1.73 (1.35)	1.65 (1.37)	0.63 (0.50)	1.07 (0.69)
+ EVA	1.13 (1.04)	1.05 (0.97)	0.76 (0.66)	0.81 (0.66)	1.75 (1.23)	1.55 (1.34)	0.57 (0.65)	0.90 (0.87)
+ SEBS	1.10 (1.05)	0.97 (0.93)	0.70 (0.77)	0.68 (0.71)	1.28 (1.15)	1.29 (1.26)	0.77 (0.71)	0.95 (0.86)
AC-20+ EVA	1.23 (1.22)	0.91 (1.04)	0.67 (0.81)	0.69 (0.92)	1.69 (1.49)	1.51 (1.60)	0.56 (0.57)	0.86 (0.92)
AC-20+ SEBS	1.19 (1.10)	1.02 (0.93)	0.45 (0.72)	0.46 (0.66)	1.66 (1.28)	1.52 (1.35)	0.59 (0.55)	0.85 (0.79)
AC-5+ EVA	1.33 (1.33)	1.01 (1.10)	0.39 (0.61)	0.39 (0.68)	1.74 (1.71)	1.45 (1.55)	0.53 (0.59)	0.79 (0.95)
AC-5+ SEBS	1.42 (1.22)	0.95 (1.00)	0.62 (0.50)	0.60 (0.50)	2.13 (1.74)	1.42 (1.62)	0.55 (0.45)	0.82 (0.72)

EVALUATION OF TENSILE STRENGTH

The indirect tensile strength is the maximum tensile stress that the specimen can withstand. This test has been used for characterizing the resistance to failure of asphalt concrete caused by tensile stresses. Another important property obtained from the indirect tensile test is the stress-strain characteristics. The tensile failure strain is usually used for predicting cracking potential. Mixes that can tolerate high strains prior to failure are more likely to resist cracking than the mix that cannot endure high strains.

The indirect tensile strengths of the unaged dense-graded and open-graded mixtures at 5 °C (41 °F) and 25 °C (77 °F) are displayed in Table 2. It can be seen that all of the modified asphalt mixtures, with the exception of those containing EVA and SEBS modified AC-5 binders, have higher tensile strength than the conventional unmodified AC-30 asphalt mixtures. In addition, the tensile strength of open-graded mixtures is lower than that of dense-graded mixtures. The indirect tensile strength of open-graded unaged mixture is approximately 135 psi, whereas the unaged dense-graded mixtures has an indirect tensile strength up to 190 psi. This is probably due to the high air void content of the open-graded mixtures. It is believed that a high air void content causes the mixtures to have higher deformation and lower strength. It can be seen that the mixtures containing modified AC-5 asphalts did not provide the same indirect tensile strength as the mixtures containing conventional AC-30 asphalt.

The rate of age-hardening of asphalt mixtures as evaluated by means of the ratios of indirect tensile strength at temperature of 25 °C (77 °F) after and before the forced-draft oven aging process for both dense- and open-graded mixtures are shown in Table 3. As seen from this table, all of the modified dense-graded asphalt mixtures with the exception of SBR modified AC-30 asphalt mixture show slower rates of age-hardening than that of conventional unmodified AC-30 asphalt mixture. The SEBS modified AC-30 mixture shows the lowest aging index as compared with the others.

The rate of age-hardening of the open-graded mixtures is slightly slower than that of dense-grade mixtures on the basis of the indirect tensile strength ratio. In addition, all of the modified open-graded asphalt mixtures, with the exception of EVA modified AC-20, AC-5 and SEBS modified AC-5 asphalt mixtures, show slower rate of age-hardening than that of conventional unmodified AC-30 asphalt mixture at temperature of 25 °C (77 °F). Again, the SEBS modified AC-30 asphalt mixture shows the lowest aging index than the others on the basis of strength ratio. All of these evidences indicate that the modifiers have the potential to slow down the rate of age-hardening of conventional asphalts. However, the forced-draft oven and the natural 6-month roof aging process did not cause a significantly higher indirect tensile strength at temperature of 5 °C (41 °F) for both dense- and open-graded modified asphalt mixtures.

EVALUATION OF FRACTURE ENERGY

The required energy to cause fracture of asphalt mixtures was evaluated by the fracture energy, which was determined by calculating the total area under the stress-strain curve from indirect tensile strength test. The typical stress-strain plots of open-graded and dense-graded mixtures at 25 °C (77 °F) are shown in Figure 1. The test results indicated that the fracture energy slightly increases after the 6-month natural roof aging, but decreases after the forced-draft oven aging for either dense-graded or open-graded mixture at both 5° and 25 °C (41° and 77 °F). This indicates that the fracture energy value tend to increase up to a maximum and then decrease as the asphalt hardens.

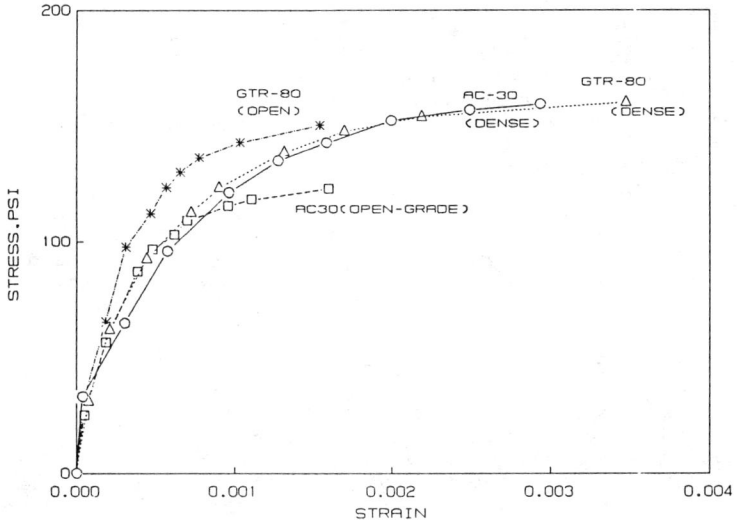

FIG. 1--Stress-strain relationship of open- and dense-graded mixtures at 25 °C (77°F)

As seen from Table 1, all of the unaged modified asphalt mixtures have higher fracture energy than that of mixtures containing the unmodified AC-30 for dense-graded mixture at 5 °C (41 °F). At temperature of 25 °C (77 °F), all of the unaged modified asphalts, with the exception of SBR modified AC-30, EVA modified AC-20 ,EVA and SEBS modified AC-5 asphalt mixtures, show higher fracture energy value than that of the conventional unmodified AC-30 mixtures. However, all of modified asphalt mixtures decrease in fracture energy after the forced-draft oven aging, for either dense-graded or open-graded mixtures at both 5 °C (41 °F) and 25 °C (77 °F). This indicates that the addition of certain amount of modifiers could increase the fracture energy of conventional asphalt. However, from the standpoint of long-term aging characteristic, the addition of modifiers to conventional AC-30 tends to make the binders too brittle as seen from these results. All of these evidences indicate that the modified AC-30 binders are generally too hard, especially after they harden further under service. However, the modified AC-5 binders appear to be softer than the conventional AC-30 as seen from these analysis results.

PREDICTION OF LOW TEMPERATURE PROPERTIES

The total resilient modulus at 25 °C (77 °F) were plotted against the corresponding total resilient modulus at 5 °C (41 °F) to evaluate their relationships. Figure 2 shows the relationship between the total resilient modulus at 5 °C and that at 25 °C for both dense-graded and open-graded mixtures. The relationship can be defined by the linear equation

$$\text{Log}(Mr5C) = \beta_0 + \beta_1 \text{Log}(Mr25C) \tag{1}$$

where

Mr5C	=	total resilient modulus at 5 °C, psi
Mr25C	=	total resilient modulus at 25 °C, psi
β_0, β_1	=	regression coefficients

Results of regression analysis indicate that the total resilient modulus at 25 °C (77 °F) has a good correlation with the total resilient modulus at 5 °C (41 °F). R^2 values of 0.80 and 0.84 were obtained for both dense-graded and open-graded mixtures, respectively. This relationship indicates that the data at temperature of 25 °C (77 °F) could be adequately considered as predictors of low temperature data.

A similar linear model was used to best-fit the relationship between the indirect tensile strength and total resilient modulus at 25 °C (77 °F) for both dense-graded and open-graded modified and unmodified asphalt concrete mixtures. The relationships between these two properties are shown in Figure 3. The equation used was of the following form:

$$\text{Log}(Mr) = \beta_0 + \beta_1 \text{Log}(St) \tag{2}$$

where

Mr	=	total resilient modulus at 25 °C (77 °F), psi
St	=	indirect tensile strength at 25 °C (77 °F), psi
β_0, β_1	=	regression coefficients

As seen from Figure 3, the best fitted equation gave a linear relationship with a R^2 value of 0.88 for dense-graded mixtures and 0.83 for open-graded mixtures. The relationship between indirect tensile strength and total resilient modulus appears to be similar between all modified asphalt concretes even though significantly different air voids

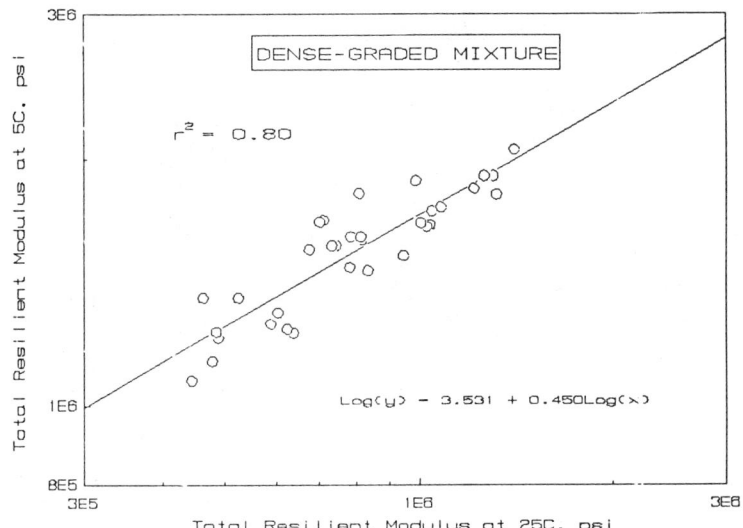

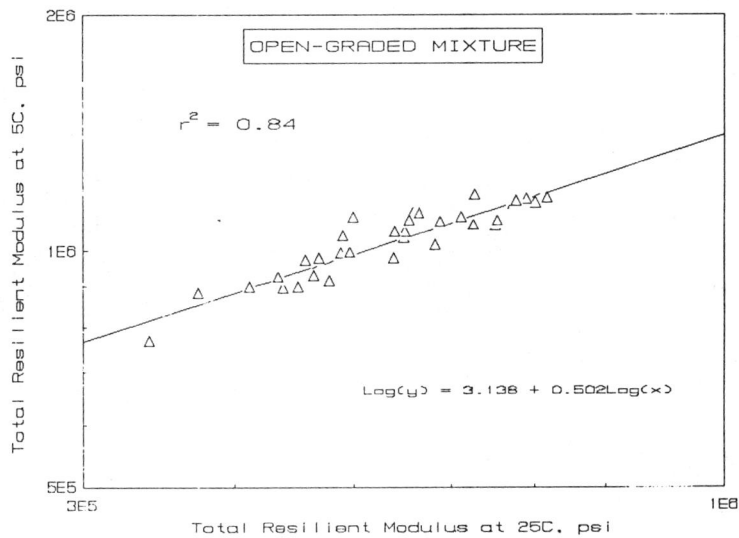

FIG. 2--Relationship between total resilient modulus at 5°C and that of 25°C

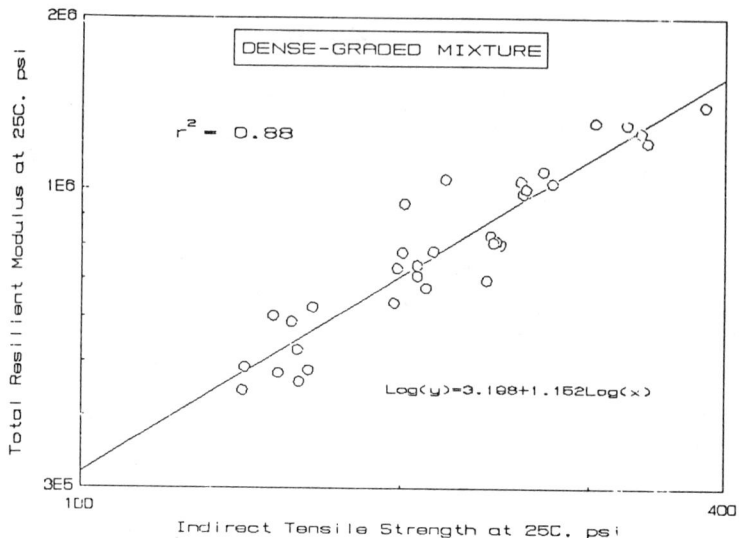
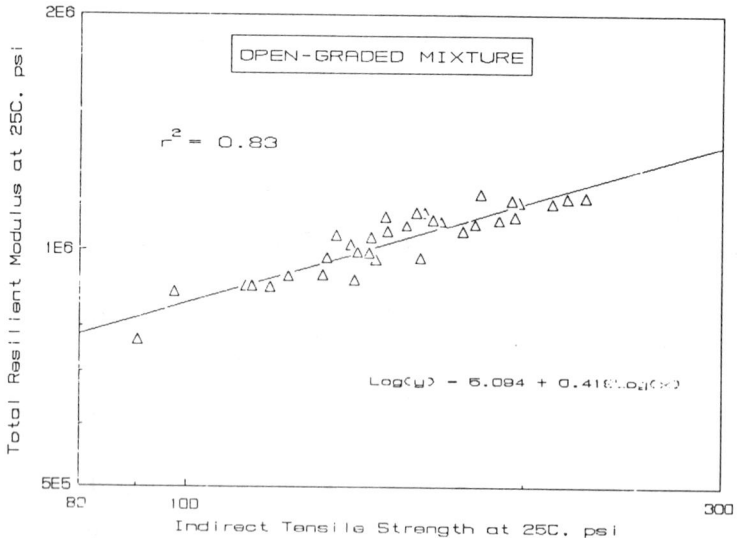

FIG. 3—Relationship between total resilient modulus and indirect tensile strength at 25 °C

were obtained from each mix. This suggests that the relationship may be independent of mixture types. In general, the total resilient modulus increases with an increase of indirect tensile strength at 25°C (77°F).

The total resilient modulus at 25 °C (77 °F) was plotted against the corresponding fracture energy at 25 °C in Figure 4. The fracture energy is determined by calculating the area under strain-stress curve from the indirect tensile test. As seen from Figure 4, the total resilient modulus had poor relationship with the fracture energy. However, a general trend can still be observed from this plot. The fracture energy increases as the total resilient modulus increases until the resilient modulus reaches a maximum around 8×10^5 psi for dense-graded mixtures and 5×10^5 psi for open-graded mixtures. As the total resilient modulus increases beyond that point, the fracture energy value tends to decrease. Therefore it is concluded that the resilient modulus test does not provide a good indication of low temperature cracking potential from the standpoint of fracture energy.

ASSESSMENT OF CRACKING POTENTIAL

An asphalt mixture will crack in service when the load- and/or temperature-induced stresses or strains exceed the failure stress or strain of the mixture under the particular condition. One of criteria commonly used in the evaluation of potential cracking of an asphalt pavement is the horizontal strain at the bottom of the bound surface layer. When the induced horizontal strains at the bottom of the surface layer exceed the tensile failure strain of the surface layer material, the asphalt mixture will crack. If the surface material has a tensile failure strain exceeding the anticipated induced strain under the worst possible condition, the material would be less likely to crack.

The cracking potential of the modified asphalt mixtures used in this study was assessed by comparing their tensile failure strains with the maximum anticipated strains if they were to be used as a surface course material in a typical Florida highway pavement. Figure 5 shows the typical Florida pavement section used in this analysis. The BISAR elastic layer computer program [2,3] was used to compute the maximum horizontal strain at the bottom of the asphalt layer caused by a 12-kip wheel load. This represents the wheel load from a worst scenario of a 24-kip single axle load. The maximum load-induced tensile strains at the bottom of the asphalt layer were computed for different values of elastic modulus of the asphalt layer material while other pavement parameters were kept constant. Figure 6 shows the relationship between the maximum tensile strain and the elastic modulus of the surface material in this hypothetical pavement system. As can be seen from the plot, as the resilient modulus of the asphalt mixture increases due to lower temperature or age-hardening, the load-induced strain also decreases. However, as the resilient modulus of the material increases, the failure strain also decreases. A lower failure strain means that the material may be more likely to crack.

The relationship between the failure strains and resilient modulus of the modified asphalt mixtures evaluated in this study are also plotted on Figure 6 for comparison. It can be seen that as the resilient modulus increases, the failure strains move closer towards the anticipated maximum strains, and thus the mixtures are more likely to crack. It can be seen that, with the exception of those mixtures containing EVA and SEBS modified AC-5 asphalts, the modified asphalt mixtures show higher failure strains than the conventional AC-30 asphalt mixtures at comparable values of resilient modulus. This indicates that the proper amount of certain modifiers to the base asphalt mixtures can increase the failure strain and therefore reduce the cracking potential.

140 ENGINEERING PROPERTIES OF ASPHALT MIXTURES

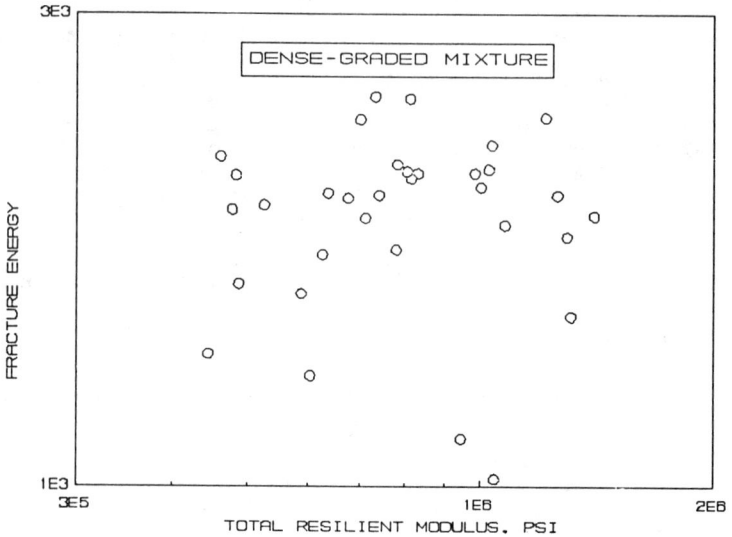

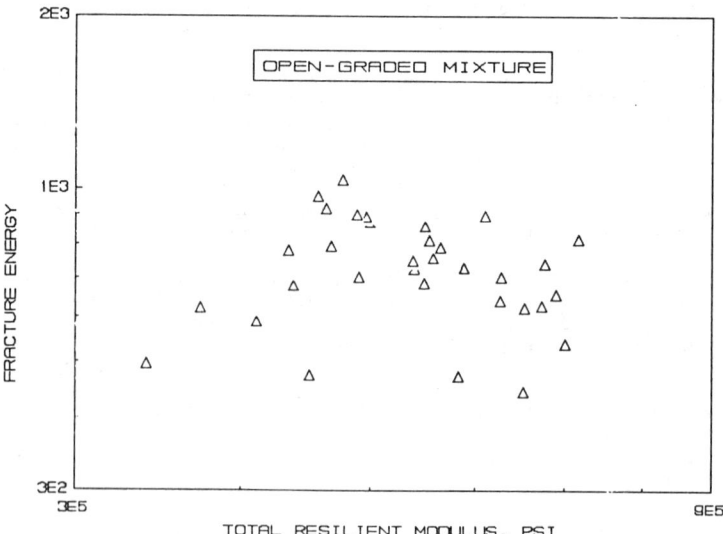

FIG. 4--Relationship between total resilient modulus and fracture energy at 25 °C

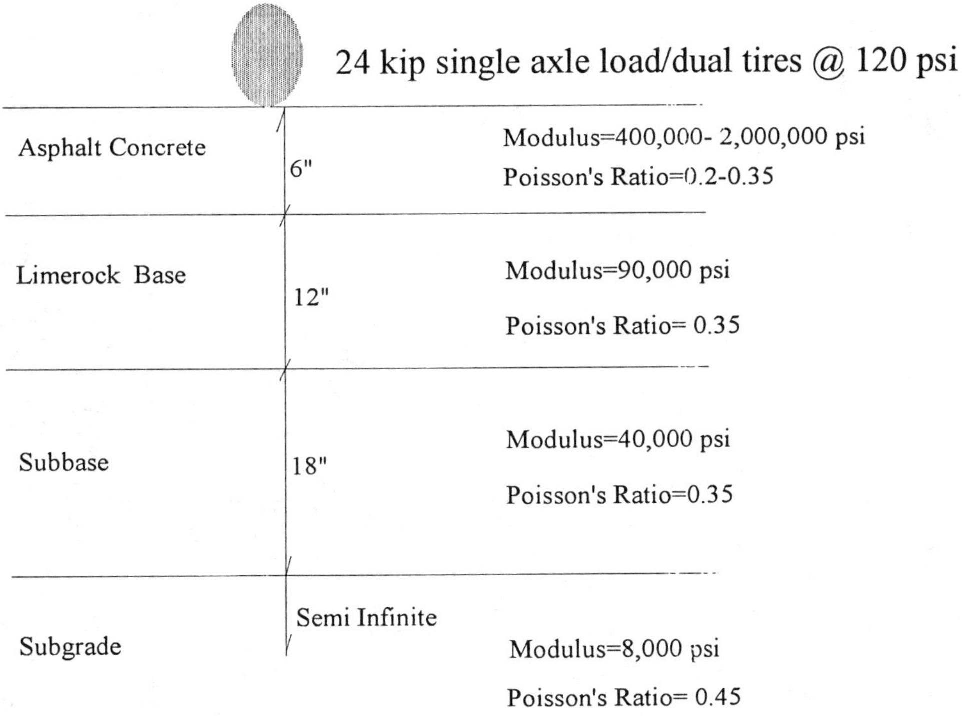

 24 kip single axle load/dual tires @ 120 psi

Layer	Thickness	Properties
Asphalt Concrete	6"	Modulus=400,000- 2,000,000 psi; Poisson's Ratio=0.2-0.35
Limerock Base	12"	Modulus=90,000 psi; Poisson's Ratio= 0.35
Subbase	18"	Modulus=40,000 psi; Poisson's Ratio=0.35
Subgrade	Semi Infinite	Modulus=8,000 psi; Poisson's Ratio= 0.45

FIG. 5   Typical pavement section for BISAR analysis

142 ENGINEERING PROPERTIES OF ASPHALT MIXTURES

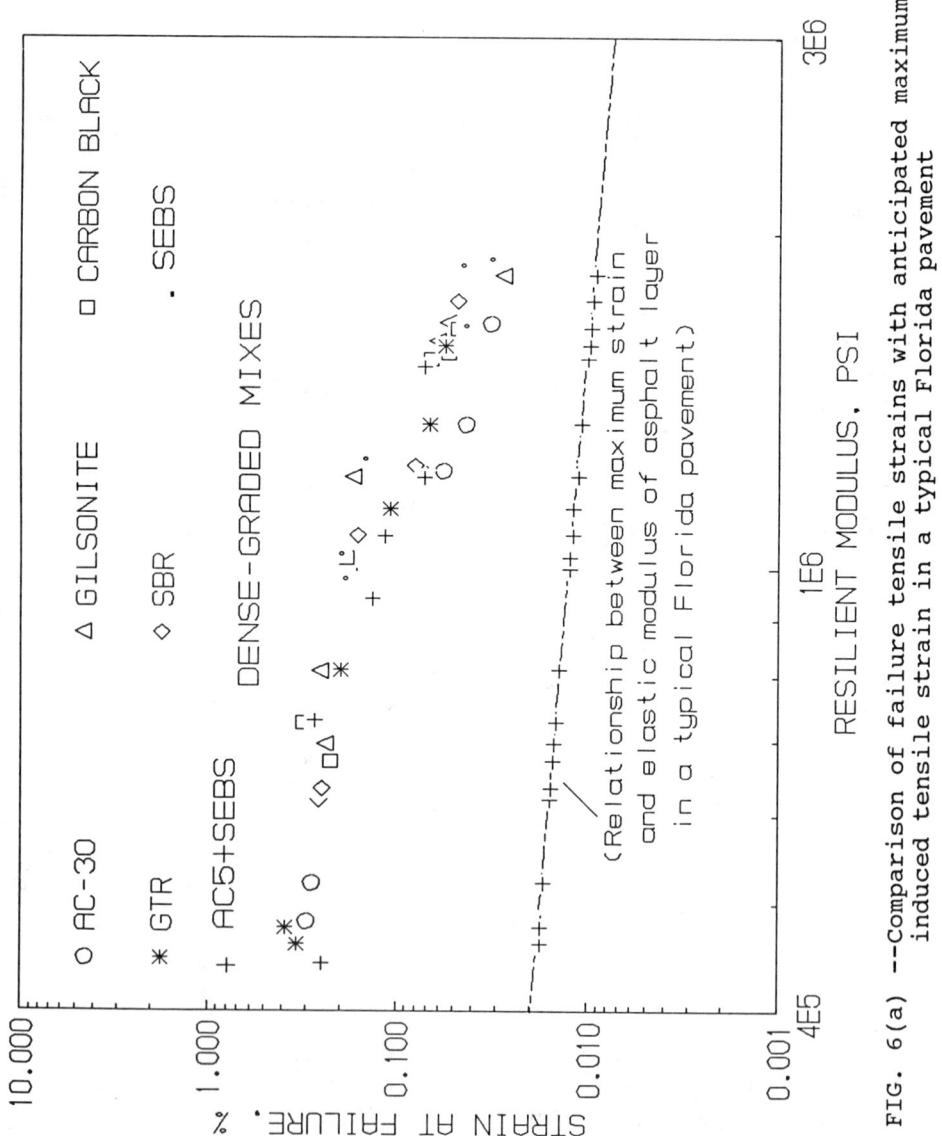

FIG. 6(a) --Comparison of failure tensile strains with anticipated maximum induced tensile strain in a typical Florida pavement

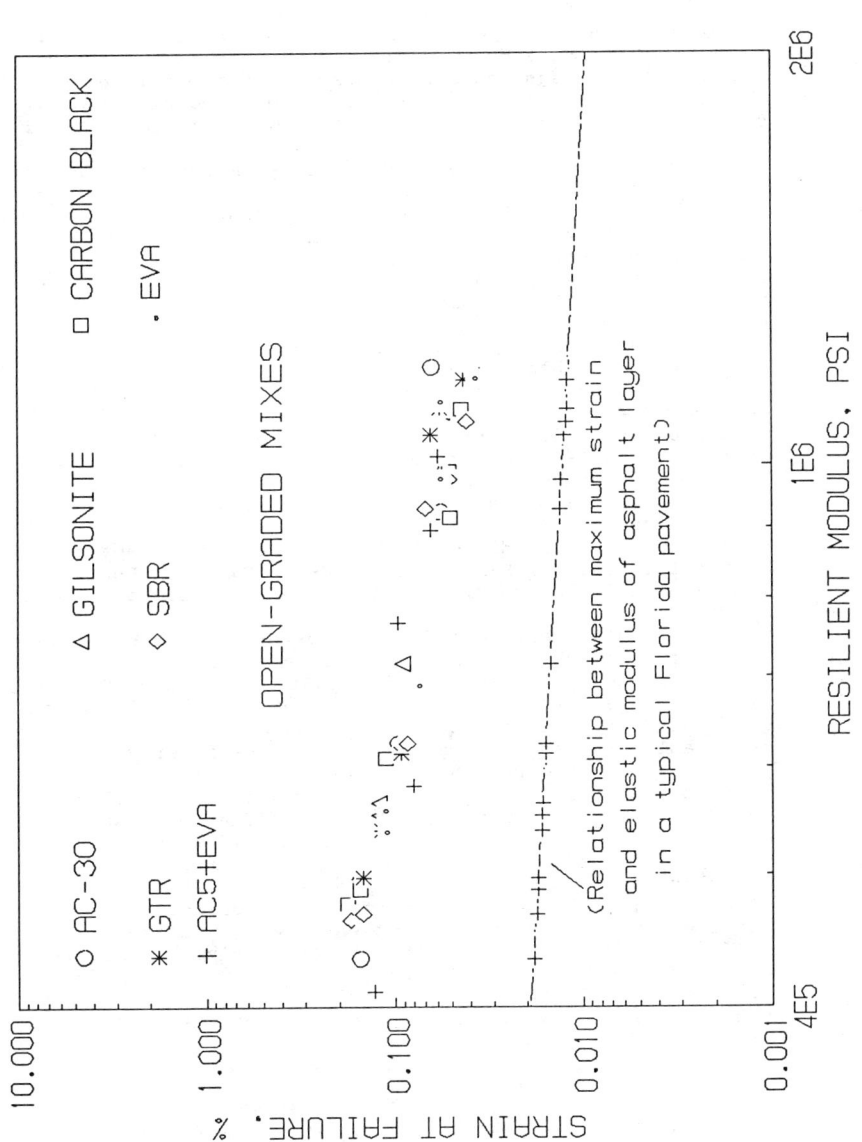

FIG. 6(b) —Comparison of failure tensile strains with anticipated maximum induced tensile strain in a typical Florida pavement

CONCLUSIONS

The major findings from the results of the mixture studies are summarized as follows:
1. The resilient modulus of modified asphalt mixtures at temperature of 25 °C (77 °F) appears to increase as the cycle frequency increases, However, at temperature of 5 °C (41 °F), the effect of test cycle frequency is not significant for either the unmodified or modified asphalt mixtures. The increase in resilient modulus as the cycle frequency increases could be explained by the delayed elastic response of the material. The delayed elastic response is more prominent at 25 °C and is negligible at 5 °C.
2. The rate of age-hardening of open-graded mixtures is slower than that of dense-graded mixtures as measured by means of resilient modulus ratio and tensile strength ratio. This may be due to the greater binder film thickness in the open-graded mixtures.
3. The fracture energy, which is characterized by means of the area of stress-strain curve increases after 6-month natural roof aging (which simulates 1-2 years of field aging) and then decreases after force-draft oven aging at 60 °C for 90 days (which simulates 3-6 years of field aging) for either dense-graded or open-graded mixtures at both 5 °C (41 °F) and 25 °C (77 °F).
4. On the basis of fracture energy, the addition of modifiers to AC-30 tend to make the binders too brittle as they harden further in service. However, the addition of modifiers to AC-5 tend to make the binders softer than the conventional AC-30.
5. All of the unaged modified asphalts, with the exception of SBR modified AC-30, EVA modified AC-20, EVA and SEBS modified AC-5 are superior to AC-30 on the basis of fracture energy.
6. The modified asphalts which are superior to AC-30 on the basis of aging characteristics include (1) AC-20 + 3.5% SEBS, (2) AC-30 + 5% SEBS, and (3) AC-30 + 5% GTR-80.
7. Since the diametral resilient moduli at 25 °C (77 °F) had a good correlation with the moduli at 5 °C (41 °F), it appears reasonable that the data at 25 °C could be considered as predictor of the corresponding low temperature data.
8. Different mixture types appear to have different aging characteristics as evidenced by the difference between the open and dense graded mixtures used in this study.
9. The resilient modulus at 25 °C (77 °F) had a good correlation with the indirect tensile strength at 25 °C (77 °F). In general, the resilient modulus increases with an increase in indirect tensile strength at a particular loading rate. In addition, this relationship does not vary with different types of mixtures.
10. GTR and SEBS modified asphalt mixtures appear to have lower rate of age-hardening than that of conventional asphalt mixtures.

REFERENCES

[1] Commandur T. Chari, Byron E. Ruth, Mang Tia and Gale C. Page, "Evaluation of Age Hardening on Asphalts and Mixtures," <u>Journal of the Association of Asphalt Paving</u>

Technologists, Vol. 59, 1990, pp176-239.

[2] Ruth, B.E., M. Tia, and K. Badu-Tweneboah, "Structural Characterization and Stress Analysis of Flexible Pavement System," Final Report to FDOT, State Project No. 99700-7351-010, UF Project No. 4910450424512, Dept. of Civil Engineering, University of Florida, Gainesville, Florida, July, 1987.

[3] Roque, R., and B.E. Ruth, "Materials Characterization and Response of Flexible Pavements at Low Temperatures," Proceedings of the Association of Asphalt Paving Technologists, Vol. 56, 1987, pp130-167.

Peter Sebaaly[1], Venu Thirmarayappa[2], Jon Epps[3], and Ledo Quilici[4]

ANALYSIS OF A WELL-PERFORMING DESERT PAVEMENT

REFERENCE: Sebaaly, P., Thirmarayappa, V., Epps, J., and Quilici, L., **"Analysis of a Well-Performing Desert Pavement,"** Engineering Properties of Asphalt Mixtures and the Relationship to their Performance, ASTM STP 1265, Gerald A. Huber and Dale S. Decker, Eds., American Society for Testing and Materials, Philadelphia, 1995.

ABSTRACT: A flexible pavement project located in the southern part of the state on I-15 near Las Vegas was selected for this investigation. The pavement section consists of an overlay that was constructed in 1970. Twenty two years later, this pavement is still showing excellent performance under the severe weather conditions and heavy traffic loading. As part of this investigation, slabs and cores were cut from the pavement section and laboratory tests were conducted on the various materials. The rheological properties of the recovered asphalt binder were evaluated and compared to the newly developed SHRP specifications. The properties of mixtures that were evaluated included the following: In situ air voids, resilient modulus at various temperatures, moisture sensitivity, and permanent deformation characteristics. The extracted aggregates were also tested for any degradation and their gradation curves were compared to the FHWA and SUPERPAVE recommendations.

KEYWORDS: asphalt binder, dust ratio, SUPERPAVE, SHRP, rutting, temperature susceptibility, and rheological properties.

INTRODUCTION

The performance of hot mix asphalt pavement is largely dependant upon the adequacy or quality of mixture design, structural design and the construction operation. Mixture design involves the selection and proportioning of materials to meet acceptance criteria. Structural design involves the determination of the thicknesses of structural layers (hot mix asphalt, base course, subbase course, etc.) to withstand the anticipated traffic for the climate (temperature, moisture etc) in which the pavement is placed and the subgrade or natural soil upon which the pavement is placed. The construction operation is responsible for physically preparing and placing the subgrade and the structural pavements layers to meet the specifications for the project. These specifications typically establish levels of quality (strength), thickness, smoothness, and uniformity.

---

[1] Associate Professor, Department of Civil Engineering, University of Nevada, Reno, NV 89557.

[2] Graduate Research Assistant, Department of Civil Engineering, University of Nevada, Reno, NV 89557.

[3] Professor, Department of Civil Engineering, University of Nevada, Reno, NV 89557.

[4] Bituminous Materials Engineer, Nevada Department of Transportation, Carson City, NV 89712.

The acceptance criteria for mixture design, pavement design and construction quality has been established over the years based on relationships between test methods which measure properties of materials; levels of stress, strain and/or deflection in the structural section; and measures of pavement performance such as rutting, cracking, bleeding, raveling, etc. These criteria have changed over the decades as traffic increased, new materials and research advancement became available. The performance data base upon which these criteria are based is limited, variable and not well documented. Thus, research to more accurately define acceptance criteria has been recognized as important and is a major part of the Strategic Highway Research Program (SHRP) and subsequent Federal Highway Administration (FHWA) research program and the American Association of State Highway and Transportation Officials (AASHTO) research program.

The research project described in this paper contributes to the body of knowledge necessary to improve the acceptance criteria for hot mix asphalt mixtures. An interstate highway pavement in southern Nevada with excellent performance over a 20 year period was selected, sampled and tested. Tests performed included routine materials and mixture tests as well as newer tests which are used for research purposes and are part of the SHRP SUPERPAVE approach to hot mix asphalt design.

RESEARCH APPROACHES

One way to establish relationship between pavement performance and materials properties is to measure properties of hot mix asphalt obtained from existing pavements with performance history ranging from very good to very poor. The post construction study is also costly and requires a large data base and thus numerous pavements must be sampled, tested and evaluated. In addition the properties of samples taken from existing pavements (field-mixed, field-compacted and field-aged) must be correlated with the properties of laboratory-mixed, laboratory-compacted, and laboratory-aged samples which are used for mixture design purposes. It is also desirable to establish correlations with field mixed, laboratory compacted samples obtained during construction to establish a construction quality control criteria.

PROJECT DESCRIPTION

As mentioned earlier the pavement section investigated here is on an interstate route which was overlaid under NDOT contract number 1392. The project is located on I-15 between 3.8 km north of the junction with US-93 and 2.0 km north of the junction with SR-40 in Clark County. There were a number of different typical sections on the project, which were required to correct the existing structural problems, but the overlay portion of the project was a 25 mm leveling course with 38 mm of plantmix surfacing and an open-graded wearing course. The project was completed in January of 1972.

The overlay was designed for a total of 1.6 million 80 kN equivalent single axle loads (ESAL). Since completion, the project has realized a total of 8.5 million 80 kN ESALs. Performance has far exceeded the design life of the overlay.

The paving contractor was Las Vegas Paving, Inc. The contractor used a Madsen 6000 pound batch plant. The material was produced using Douglas 60-70 Penetration grade asphalt cement. The aggregate gradation consisted of 100% passing the 12.5 mm sieve and 100% crushed faces.

In Southern Nevada, it was not unusual for NDOT to design bituminous mixes with 12.5 mm aggregate. It was also typical to design mixes with air voids in the range of 7 to 9 percent range.

EXPERIMENTAL DESIGN

In November, 1992, the Nevada Department of Transportation (NDOT) obtained samples from the I-15 section (Clark County, NV) for the laboratory investigation. A total of 30 cores and 6 slabs were obtained. Fifteen cores and 3 slabs were obtained from each of the north and south bound directions. The cores were 102 mm in diameter and 204 mm in average height. The slabs were 305 mm wide, 1220 mm long, and 204 mm in average depth. The samples contained both the original asphalt concrete layer built in 1959 and the asphalt concrete overlay built in 1972. The samples from both layers were tested in this investigation, however, the properties of the overlay layer were of greater interest to this research. The maintenance data was available from 1980 to 1991. The PMS data was provided by NDOT from 1983 onwards.

Figure 1 shows the flow diagram of the testing program that was conducted. Twelve cores from each direction were separated for top and bottom layers. The cores were cut into 63.5 mm height and were used to test for the resilient modulus temperature series and the moisture sensitivity series. Samples were taken from the slabs for extraction and subsequent testing of the asphalt cement and aggregates. Two cores (full depth) from each direction were used to test for permanent deformation analysis.

MIXTURES DATA

Air Voids

The air voids were measured for cores from both directions using the rice specific gravity and the saturated surface dry bulk specific gravity. Table 1 summarizes the density and air voids data. The air voids ranged from 6.7 to 8.8 percent for both directions. Pavement performance studies have shown that in-place air voids greater than 3 percent are necessary to decrease the probability of premature rutting throughout the life of the pavement (1). An in-place air voids less than 2.5 percent greatly increases the probability of premature rutting. The data from this pavement section fall well above these recommendations.

Table 1. Summary of air-voids analysis.

Direction	Sample number	Specific gravity (SSD)	Ther. max. sp. gr.	Air voids
South bound	1	2.278	2.467	7.66
	2	2.237	2.441	8.35
North bound	1	2.289	2.458	6.87
	2	2.296	2.461	6.70

Resilient modulus Data

The resilient modulus test was used to determine the elastic characteristics of the asphalt concrete mixture. The AASHTO design guide recommends that ASTM D-4123 be used to measure the resilient modulus of asphalt concrete mixtures for the use in structural design (2). Resilient modulus at 25°C was measured for twelve cores for each direction. The M_r data were used to evaluate the variability among the cores before

SEBAALY ET AL. ON DESERT PAVEMENT   149

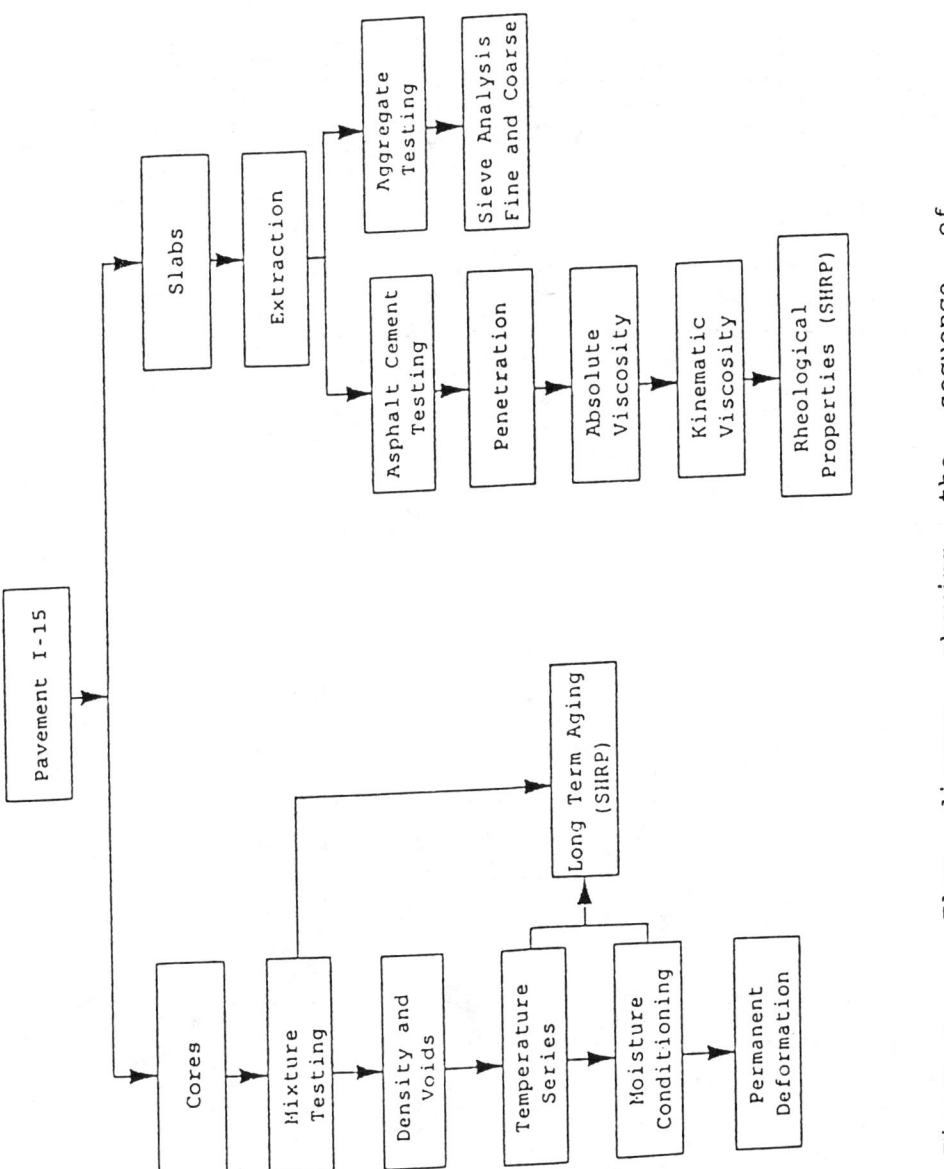

Figure 1. Flow diagram showing the sequence of laboratory testing

additional tests were conducted. Table 2 summarizes the M_r (25⁰C) for both directions. For each set, the mean, standard deviation and coefficient of variations (CV) were calculated. Based on the low coefficient of variation values in Table 2 (i.e., CV's below 10 percent) it can be seen that all of the cores from each direction have very similar M_r (25⁰C) values. Therefore, any sets of cores can be selected for further evaluation without a great concern about the representative sampling technique. Comparing the mean values from each direction, it can be seen that the southbound cores have M_r (25⁰C) values higher than the northbound cores. Therefore, it was decided to evaluate each direction separately.

Table 2. Resilient modulus values at 25⁰C for the cores.

Sample number	North bound	South bound
1	7400	11300
2	7800	11900
3	7400	11800
4	6200	9900
5	7700	9400
6	6500	11400
7	7400	13200
8	5700	13200
9	6600	12500
10	7300	12600
11	8000	12000
12	6900	12400
Mean	7000	11800
Std dev	689	1200
Coeff of var	9.84	10.20

* All M_R values in Mpa

Temperature Series Data: In this test three cores were randomly selected from each direction and their resilient modulus values were measured at 0, 25 and 40⁰C (each core was tested at all three temperatures). By looking at the modulus versus temperature curves in figures 2 and 3, it can be seen from the data that the cores have excellent temperature susceptibility characteristics. They relatively have high modulus values at higher temperatures and low modulus values at lower temperatures. These mixtures will have good resistance to low temperature cracking, fatigue cracking, and rutting which is the behavior of an ideal asphalt concrete mixture.

Comparison of Cores with New Mixtures: As indicated earlier, the temperature susceptibility of the cores was excellent. It was hypothesized that this could be due to the aging of the binder or the good properties of the initial mixture. In order to evaluate the source of this good temperature susceptibility, a new mixture was prepared using the

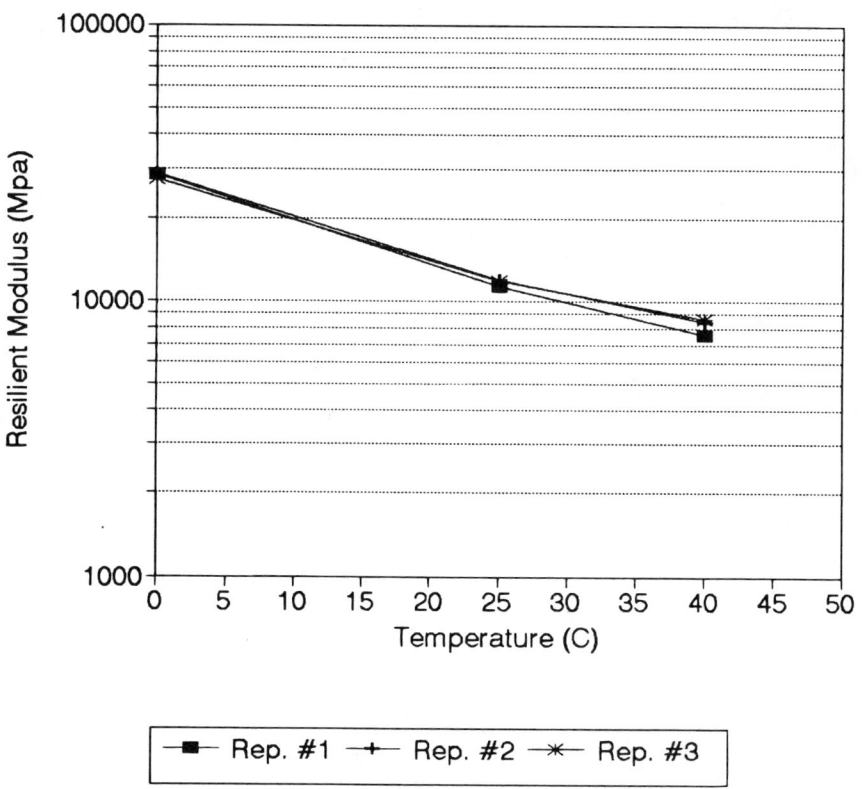

Figure 2. Resilient modulus as a function of temperature for the south bound direction.

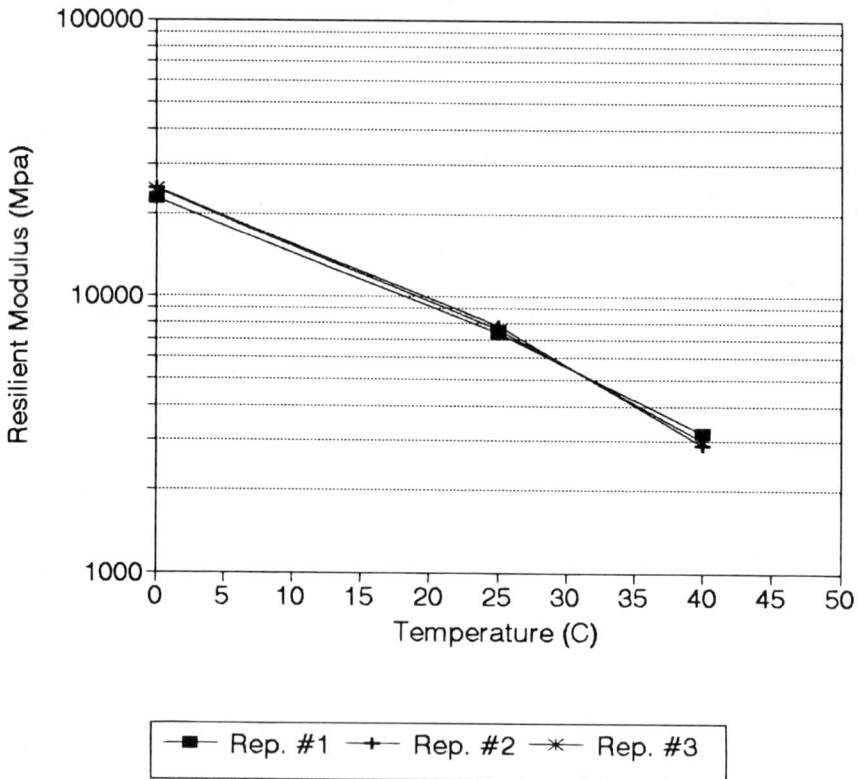

Figure 3. Resilient modulus as a function of temperature for the north bound direction.

same aggregate source and gradation along with AC-20 Witco binder. The Hveem mix design procedure was used to select the optimum AC content for these mixtures ($AC_{opt}=4\%$). Following the mix design, the resilient modulus of the new mixture were evaluated at 0, 25 and 40^0C. Next, the new mixture was aged according to the SHRP long term aging procedure (SHRP #1025) and the resilient modulus of the aged mixture were evaluated at 0, 25, and 40^0C (3).

Plotting the resilient modulus as a function of temperature for the unaged and aged mixtures as shown in Figure 4, would indicate the effect of aging on the temperature susceptibility of the mixture. It was anticipated that by aging the mixture, the temperature susceptibility curve could be either rotated or shifted upward. The data in Figure 4 indicate that aging of the mixture has shifted the temperature susceptibility curve instead of rotating it. By aging the mixtures, their resilient modulus have increased at all temperatures.

By comparing the temperature susceptibility curves for the new unaged and aged mixtures with the curves of the cores, it can be seen that the cores have better temperature susceptibility characteristics in all cases. Since the aging of the mixtures did not rotate the curve, it can be concluded that the initial mixtures used on this pavement section had excellent temperature susceptibility characteristics.

This data also indicate that the new unaged mixture with AC-20 Witco has a relatively high resilient modulus at 25^0C of 3,400 Mpa. This level of resilient modulus at 25^0C is considered sufficient and is anticipated to provide excellent long term pavement performance. Based on this data, it can be concluded that the aggregate source and gradation have played a major role in providing the excellent performance history of this pavement.

Moisture Sensitivity Data: Three cores were randomly selected from each direction for the moisture conditioning test series. The resilient modulus test was conducted at 25^0C for the cores and M_r values were measured before and after moisture conditioning. Tables 3 and 4 summarize the data for all the cores in both directions. From the data it can be observed that the mixtures in both directions have excellent retained strength ratios. This indicates that these mixtures have good resistance to moisture damage and stripping and it has shown in terms of the excellent long term pavement performance.

Table 3.   Moisture sensitivity data in the south bound direction.

Sample No.	1S7	1S8	1S9	Mean	Std dev	Coeff of var
M_R (25^0C)	13200	13200	12500	13000	400	3.0
M_R (25^0C) (Freeze /Thaw)	8600	8100	7700	8100	420	5.2
Ratio (%)	65	61	62			

* All M_R values in Mpa.

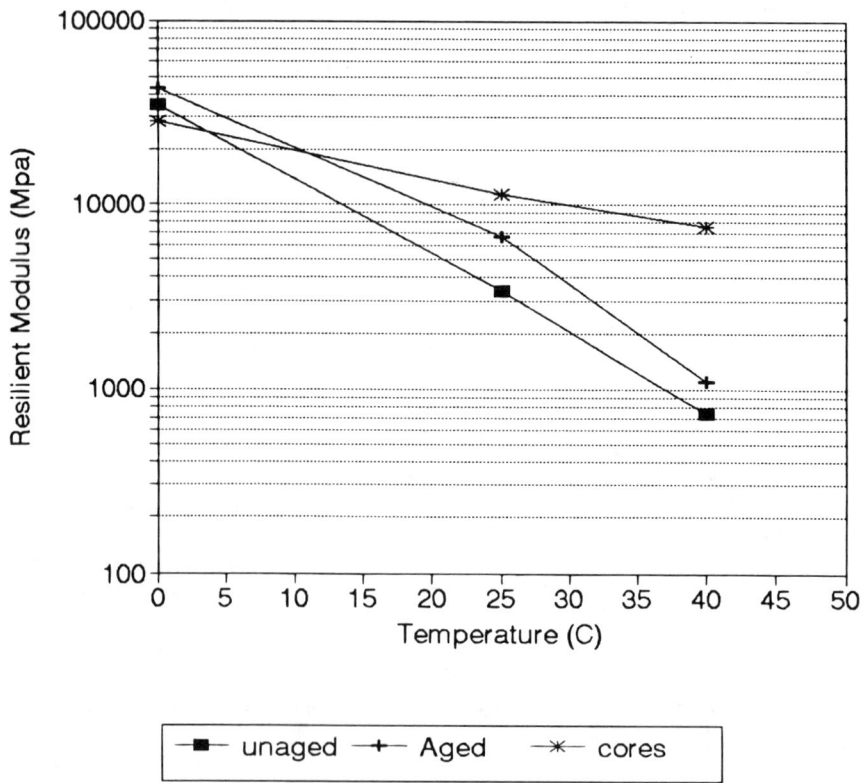

Figure 4. Comparison of unaged and aged mixtures with cores.

Table 4. Moisure sensivity data in the north bound direction.

Sample No.	1N4	1N6	1N9	Mean	Std dev	Coeff of var
M_R (25°C)	6200	6500	6600	6400	210	3.3
M_R (25°C) (Freeze /Thaw)	7400	6600	6500	6800	525	7.7
Ratio (%)	120	102	98	106		

* All M_R values in Mpa.

Tensile Strength Data

The indirect tensile test was conducted on three cores from each layer in each direction to evaluate the tensile strengths of the mixtures at 25°C. The tensile strength data are summarized in Table 5. The wet tensile strength values for all cores ranged from 1.0 to 1.6 Mpa while the dry tensile strength values ranged from 1.6 to 1.9 Mpa. These kind of tensile strength values are considered very high for asphalt concrete mixtures. Thus these mixtures have excellent tensile strengths which would provide excellent resistance to fatigue and low temperature cracking. The excellent fatigue resistance of these mixtures has drastically reduced the fatigue failure of the section. With the section being in the southern part of the state, the low temperature cracking was not a problem.

Table 5. Tensile strength data.

Direction/ Condition	Tensile strength in Mpa			Mean	Std dev	Coeff of var %
	Sample number					
	1	2	3			
South/Dry	1.6	1.7	1.6	1.6	.06	3.80
South/Wet	1.1	1.4	1.1	1.2	.20	16.7
North/Dry	1.8	1.9	1.9	1.9	.07	3.70
North/Wet	1.2	1.6	1.2	1.3	.21	16.1

* All Tensile Strength Data in Mpa.

Permanent Deformation Data

Rutting is caused by permanent deformation in any or all of the pavement layers and subgrade, it is usually caused by consolidation or lateral movement of the material due to traffic loads.

The PMS data for the rut depth showed that the majority of the measured rut depth values are less than 3 mm. Pavement performance studies have shown that the rut depth less than 4 mm is insignificant. This shows that this pavement is highly resistant to rutting.

156   ENGINEERING PROPERTIES OF ASPHALT MIXTURES

The air voids measurements indicate that this pavement has in-service air voids between 6 to 9 percent for the top layer. It can be seen from this data that the pavement has maintained a relatively high air voids level even after 20 years of service. As mentioned earlier, mixtures for the southern region of Nevada were designed to provide air voids in the range of 7 to 9 percent.

The permanent deformation data for both directions are summarized in Figures 5 and 6 which indicate that the plastic strains are very small. It should be noted that the permanent deformation samples consisted of both layers. The south bound layer has smaller plastic strain of about 0.16 percent and the north bound layer has a higher plastic strain between 0.4 and 0.6 percent after 12,000 load cycles which is still a relatively small deformation. This shows that the pavement was highly durable and resistant to permanent deformation. Again aggregate type and gradation are the most significant contributors to this excellent resistance to rutting.

ASPHALT BINDER DATA

The Asphalt Cement was extracted from the pavement mixture using ASTM D-2172 Method B (the Reflux Method). Several binder tests were conducted to evaluate the properties of the aged asphalt cement.

As mentioned earlier, the original asphalt was Pen Grade 60-70. The properties of the recovered asphalt cement are summarized in Table 6. It can be seen that the asphalt contents based on the extraction data are lower (3.66 and 4.26 %) than the mix design asphalt content (4.5%) for both direction. This is expected since some of the asphalt is absorbed by the aggregates. The measured penetration values of the recovered asphalt are extremely low (5 to 7) for a 60-70 grade asphalt which indicates that the has aged as expected. The measurement of the viscosities at both the 135 and 60 °C (no flow) also indicate that the binder has aged significantly.

Table 6.   Binder properties.

Binder properties		South bound	North bound
Mix Design Asphalt Content (%)		4.5	4.5
Recovered Asphalt Content (%)		3.66	4.26
Viscosity of Extracted Binder	135 °C	4884	1892
	60 °C	No Flow	No Flow
Penetration of Extracted Binder		5, 6, 7	7, 6, 7
Dust Ratio based on Extrated Data		2.35	1.90

Rheological Properties

The SHRP binder specifications call for the evaluation of the binder from the tank, after aging through the rolling thin film oven (RTFO) and after aging through the pressure aging vessel (PAV). In each stage certain limiting value applies at a specific testing temperature which is appropriate for the environmental region where the binder will be used.

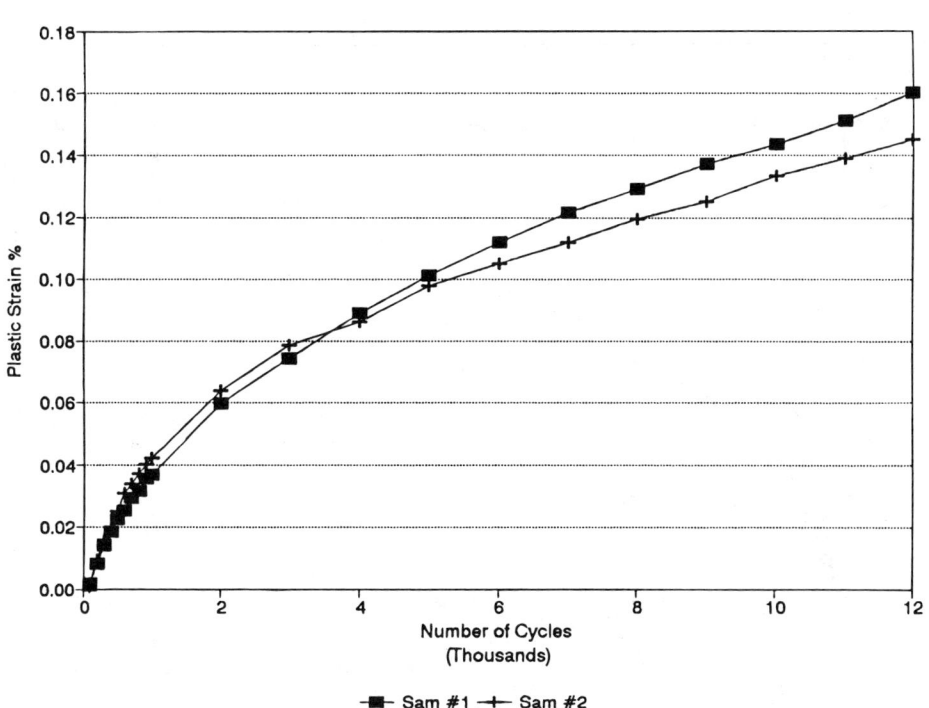

Figure 5. Permanent deformation characterstics of the south bound mixtures.

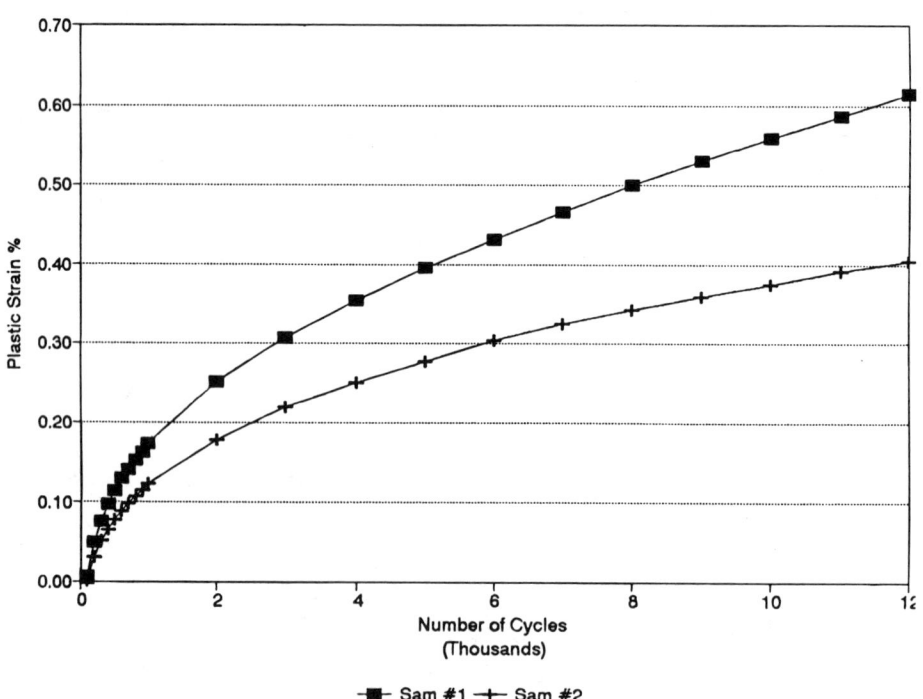

Figure 6. Permanent deformation characterstics of the north bound mixtures.

For example, the G˚Sinδ must be evaluated in order to check if the binder would meet the specifications. The specifications calls for a maximum value of G˚Sinδ as 5000 Kpa at 10 rad/sec under various levels of temperature after the binder has been aged through RTFO and PAV.

The SHRP criterion does not indicate the pavement age which is represented by the RTFO/PAV aging process. However, for the purpose of comparing this data to an established standard, this part of the research considered the extracted binder as equivalent to a binder aged through RTFO and PAV. The recovered binder was then tested on the Rheometrics Asphalt Analyzer (RAA) machine to determine its G˚Sinδ properties at temperatures of 28, 31, and 34^{0}C. The 8mm parallel plate test configuration was used at 10 rad/sec following the test procedure recommended by SHRP test method B-003 (4). Figures 7 and 8 show the G˚Sinδ as a function of temperature for the extracted binders.

The SHRP binder specification, recommends a PG76-16 binder for southern Nevada. It can be seen that the binders from both directions do not meet the SHRP specifications which calls for a maximum value of G˚Sinδ of 5,000 Kpa. According to these specifications, these binders are too brittle and will cause both fatigue and low temperature cracking. However, performance of this pavement section indicate otherwise. It should be noted that the SHRP specifications are relatively new and must undergo extensive verification studies. This research activity could be considered as one verification study for the SHRP specifications.

AGGREGATES EVALUATION DATA

Aggregates makes up 90-95 percent by weight and 75-85 percent by volume of most asphalt concrete mixtures, hence it provides most of the load bearing characteristics of the pavement. Pavement performance is heavily influenced by the choice of a proper aggregate for a particular job.

Durability Analysis

The sieve analysis of the extracted aggregates was used to determine their durability. A typical gradation of the original aggregates and the aggregates recovered from the mixture (referred to as the aged gradation) along with the NDOT specifications are shown in Figure 9. It can be seen from these curves that not much degradation has taken place in the aggregates. These aggregates are very durable and have excellent physical properties.

Dust Ratio

The dust ratio is defined as ratio of dust (passing #200) to asphalt cement content.

Dust Ratio = Dust (passing #200)/Asphalt Content

The dust ratios for the mixtures are 2.35 and 1.90 for the south and north bound respectively (Table 6). The values are based on the extracted properties. As mentioned earlier the extraction was done following ASTM D-2172 Method B which uses the reflux process. The FHWA's advisory on hot mix asphalt pavements recommends that a dust ratio between 0.6 and 1.2 should be maintained (5). The data from this pavement section indicate that the FHWA's recommendation has been violated.

SUPERPAVE Gradation

The strategic Highway Research Program has developed a mixture design and analysis system referred to as SUPERPAVE. Aggregate gradation plays a major role in this system where a restricted zone has been

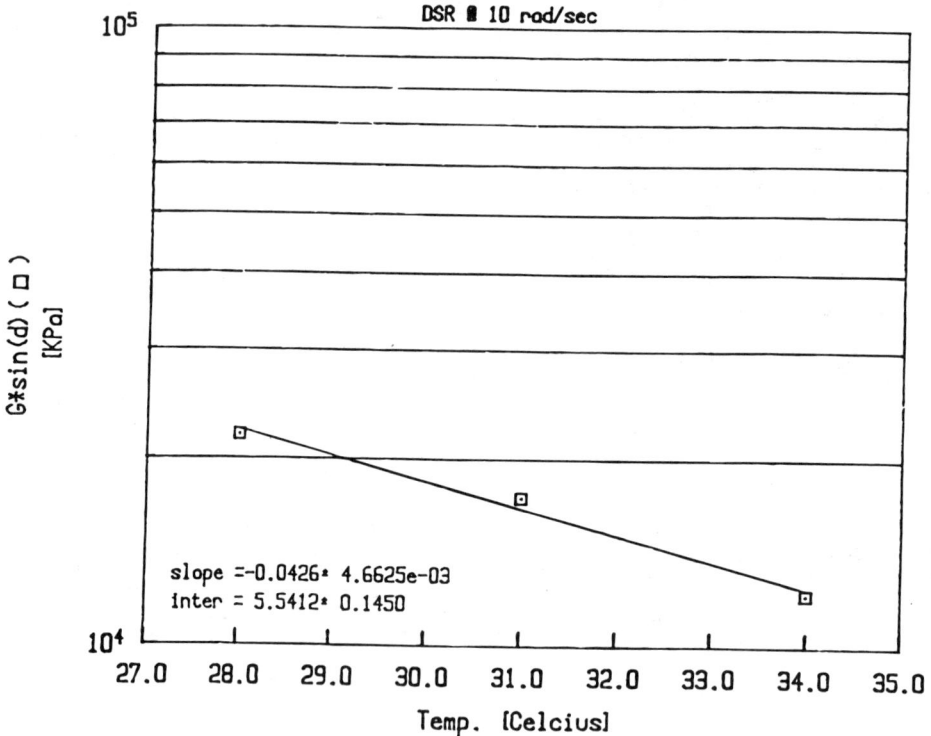

Figure 7. Temperature versus $G^*\sin\delta$ for the south bound

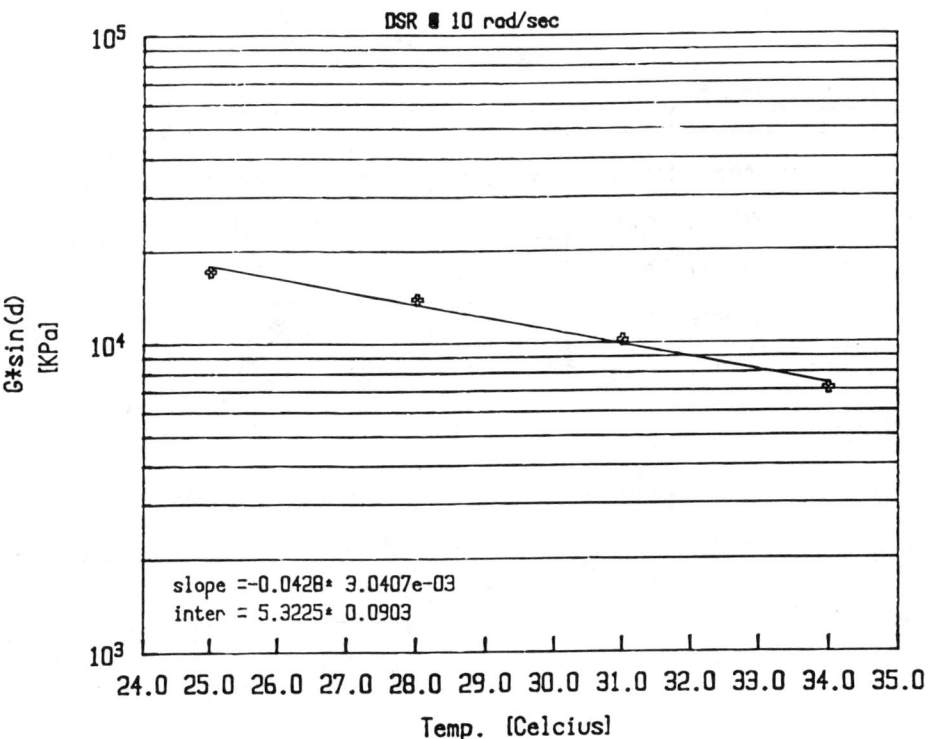

Figure 8. Temperature versus $G^*\sin\delta$ for the north bound

162 ENGINEERING PROPERTIES OF ASPHALT MIXTURES

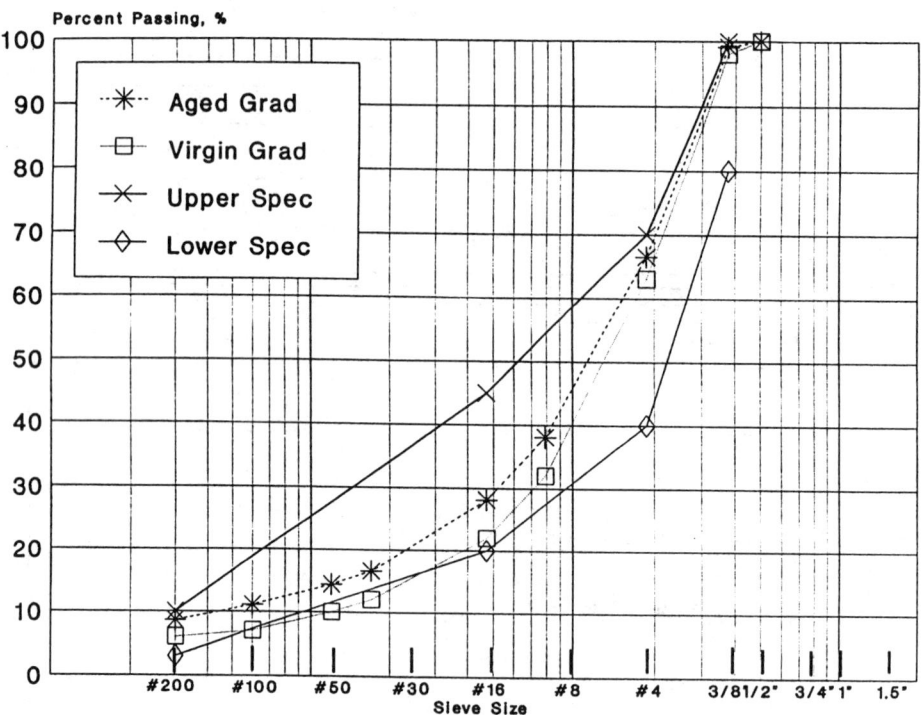

Figure 9.. Gradation of the original and recovered aggregates from the mixtures south bound

established to ensure a gradation which provides a good performing mixture. Figure 10 shows the gradation of the overlay mix plotted on the SUPERPAVE gradation chart. The forbidden zone is indicated by the darkened region on the chart. It is hypothesized in SUPERPAVE that any mixture having a gradation that crosses the forbidden zone will not lead to a good long term pavement performance.

The curves in Figure 10 show that the recovered gradation curve passes below the forbidden zone and it does not have the hump on the #40 sieve. According to SUPERPAVE recommendations, this mixture would perform very well and in reality it did.

These aggregates have 100 percent crushed materials which provides rough surface textures and angular particle shapes that enhance interparticle friction.

SUMMARY AND CONCLUSIONS

Based on the extensive laboratory evaluation that were conducted in this project it can be concluded that the pavement section has a superior asphalt concrete mixture. This superior mixture has led to excellent long term performance. The mixture, however, violates several recommendations which are used in the current practice of asphalt concrete mixture and yet still provided superior field performance. This indicates that the excellent quality materials combined with practical mix design and good construction practice can provide excellent long term performance.

The following represent a summary of the interesting observations made during the evaluation of these mixtures.

1. The variability of the mixtures after over twenty years in-service, as measured by their resilient modulus values at $25°C$, was very low. This indicates that the quality control/assurance during construction was extremely good.

2. The temperature susceptibility characteristics of the mixtures are excellent. The laboratory investigation revealed that this excellent temperature susceptibility is not due to the aging of the binder alone. It was concluded that the original mixtures had excellent temperature susceptibilities. The aggregate source and gradation combined with good binder and excellent mix design and construction practice have led to these excellent characteristics.

3. The moisture susceptibility of the mixtures is excellent which eliminated the potential of any stripping and raveling failures. The mixtures have shown high dry and wet strengths, desirable properties of any asphalt concrete mixture.

4. The permanent deformation data indicate that the mixtures are highly resistant to rutting. This was endorsed by the fact that the majority of the measured rut depth values after more than twenty years in service were below 3 mm. It should be noted here that the maximum size of aggregates was only half inch and the aggregate was 100 percent crushed. This indicates that it is not absolutely necessary to use large stone mixtures in order to obtain a rut resistant mix.

5. The standard binder tests data (i.e. penetration and viscosity) have indicated that the binder has aged. If one assumes that the SHRP's aging process is equivalent to twenty years service, than this binder would be classified as brittle. However, the aged properties of the binders did not hinder the performance of the pavement. This indicates that in order to predict the performance of the pavement, one should evaluate the entire mixture and not only the binder.

164 ENGINEERING PROPERTIES OF ASPHALT MIXTURES

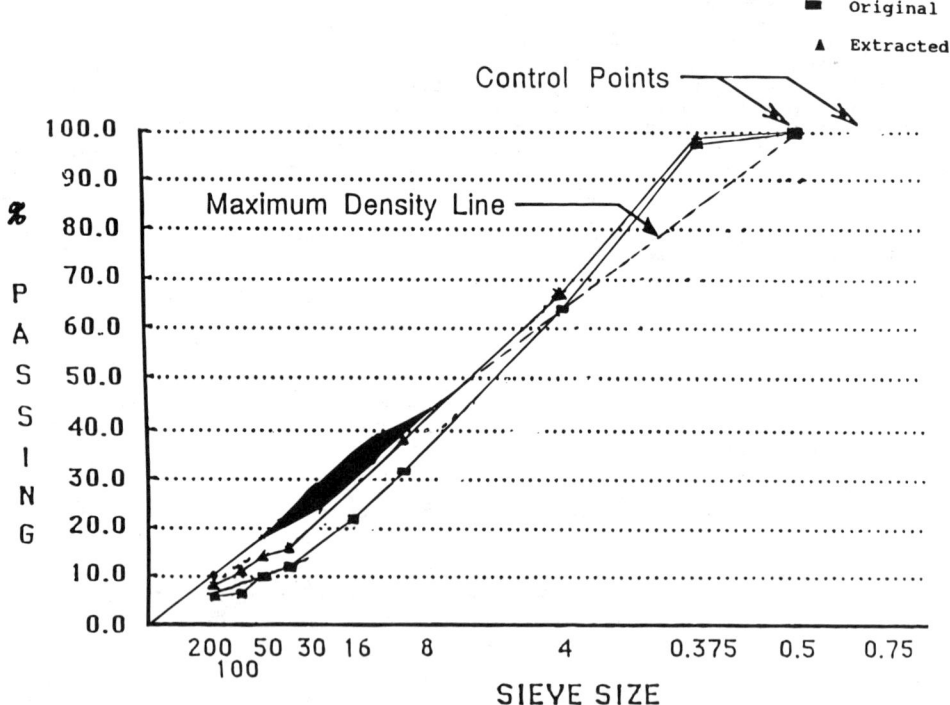

Figure 10. Aggregate gradation as related to the SUPERPAVE grading zones, south bound

6. The characteristics of the aggregate gradation were checked against several of the current standard recommendations by both FHWA and SHRP. First, the 100 percent crushed faces were maintained throughout the service life. Second, the FHWA's dust ratio recommendation was violated. The mixture had higher (almost double) the maximum recommended dust ratio. Again, this did not seem to affect the performance of this mixture. Finally, the latest recommendations on aggregate gradation, from the SHRP SUPERPAVE system, recommends that the gradation curve should stay out of a restricted zone. The mixture's gradation did indeed pass below the restricted zone.

The extensive evaluation of this pavement section showed that the combination of the following factors would provide excellent long term pavement performance eventhough the individual mixture component do not meet the commonly used standards:

- Excellent quality of aggregate.
- Aggregate gradation compatible with the source of aggregate.
- Excellent temperature susceptibility.
- Excellent moisture resistance.
- Excellent permanent deformation characteristics.

REFERENCES

1. Kumar, A., and Goetz, W.H., "Asphalt Hardening as Affected by Film Thickness, Voids and Permeability in Asphaltic Mixtures" Proceedings, The Association of Asphalt Paving Technologists, Vol 46, 1977.

2. AASHTO Guide for Design of Pavement Structures, American Association of State Highway and Transportation Officials, Washington D.C., 1986.

3. Strategic Highway Research Program (SHRP), "Long Term Aging of Asphalt Concrete Mixtures," Transportation Research Board, Washington D.C., 1993.

4. Strategic Highway Research Program (SHRP), "Performance Graded Asphalt Binder," Transportation Research Board, Washington D.C., 1993.

5. FHWA Memorandum on Asphalt Mix Design and Field Control, Washington D.C., Dec. 1987.

Yibin Li[1], and Soheil Nazarian[2]

EVALUATION OF AGING OF HOT-MIX ASPHALT USING WAVE PROPAGATION TECHNIQUES

REFERENCE: Li, Y. and Nazarian, S., "Evaluation of Aging of Hot-Mix Asphalt Using Wave Propagation Techniques," Engineering Properties of Asphalt Mixtures and the Relationship to their Performance, ASTM STP 1265, Gerald A. Huber and Dale S. Decker, Eds., American Society for Testing and Materials, Philadelphia, 1995.

ABSTRACT: A methodology for determining the degree of aging of hot-mix asphalt (HMA) is presented. The evaluation of aging involves relating the wave propagation velocities to the elastic modulus of the mixture, before and after aging. The nondestructive nature of the methodology and its high degree of repeatability make it attractive to monitor the variation in modulus of an HMA specimen through the entire aging process. In addition, field tests very similar to the laboratory tests can be easily implemented.

KEY WORDS: hot-mix asphalt, aging, laboratory testing, wave propagation, elastic modulus

The aging of hot-mix asphalt (HMA) is of increasing concern in the maintenance of flexible pavements. Therefore, an accurate quantification of the phenomenon is important.

The aging of HMA is due to factors such as volatilization, oxidation, and steric hardening of the asphalt cement in a mixture [1]. It results in an increase in the HMA's elastic modulus and brittleness. Although the increase in elastic modulus can improve the load distribution system of the pavement structure, the increase in brittleness—resulting from excessive hardening—often leads to pavement cracking [2]

---

[1] Graduate Research Assistant and [2] Assoc. Prof., Center for Geotechnical and Highway Materials Research, The University of Texas at El Paso, El Paso, TX 79968

and a loss of durability (in terms of wear resistance and moisture susceptibility) [3]. The increase continues until the HMA layer no longer responds to loading and unloading in an elastic manner.

Due to the asphalt-aggregate interaction, the aging of asphalt cement alone is not a sufficient indicator of the aging susceptibility of an asphalt-aggregate mixture [4]. The following material properties have been found to affect the rate of aging of HMA: asphalt cement characteristics [5], asphalt content and voids in the total mix (VTM) [2], aggregates [6], and environmental factors.

A variety of test methods have been used to evaluate the effects of aging of HMA mixtures. Almost all of these are comparative in nature; that is, a given parameter related to the material is measured and compared before and after some level of aging.

The ideal test procedure should be nondestructive (so that the before and after tests can be performed on the same specimen), repeatable, rapid, and easy to use. The tests of aged mixtures which are typically preferred include the following: resilient modulus test [2], fatigue test [2], creep test [7], indirect tensile test [8], dynamic modulus test, [9] and wave propagation test [10]. Of all these, the wave propagation test is considered the closest to the ideal. It is nondestructive, repeatable to within 5 percent [10], has a testing time of less than 1 minute, and allows for the simple interpretation of data.

A field device capable of performing all of the tests quite rapidly and in an automated fashion has been developed [11]. Work performed by Li [12] shows close correlation between field and laboratory wave propagation test results.

METHODOLOGY

The schematic of the testing setup is shown in Figure 1. A sonic/ultrasonic device was used. A transmitting transducer was securely placed on the top face of the specimen. This transducer was connected to the built-in high-voltage electrical pulse generator of the device. The electrical pulse transformed to mechanical vibration was coupled to the specimen. A receiving transducer was securely placed on the bottom face of the specimen, opposite the transmitting transducer. The receiving transducer, which sensed the propagating waves, was connected to an internal clock capable of automatically measuring the travel time of the compression waves. Compression waves (P-waves), which propagate along a spherical front, transmit energy through a "push-pull" motion from one point to another.

A holding device was designed and constructed to conveniently test the specimens at high temperatures. To ensure full contact between the specimen and transducer, a loading plate was placed on top of the transmitting transducer. A spring supporting system was also constructed for the bottom transducer to ensure full contact.

A thermistor thermometer was used to monitor the temperature of the specimen. The thermometer was placed inside a small 2.5-cm-deep hole drilled on the specimen 1.25 cm from its edge.

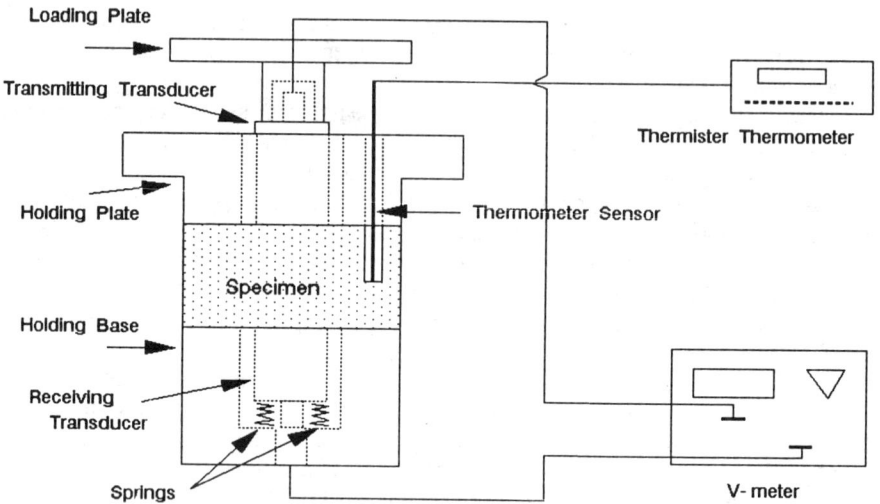

FIG. 1 — Schematic of the Test Setup

The P-wave velocity, V_p, was obtained by simply dividing the thickness of the specimen by its corresponding travel time. The elastic modulus of the specimen, E, was then calculated from

$$E = \rho V_p^2, \tag{1}$$

where ρ = mass density of the specimen.

A typical measurement would take less than one minute, and the same specimen could be tested for several aging periods. The results are detailed below.

PRESENTATION OF RESULTS

A local aggregate was utilized in this study. The HMA was Type D, a fine-graded surface course. The gradation of the HMA mixture is shown in Table 1. The asphalt cement used in this mixture was AC-20.

Specimens, 100 mm in diameter and 50 mm in height (standard briquettes), were prepared using a Texas gyratory shear compactor. Three nominal asphalt contents of 4, 5, and 6 percent were selected for the mixture. The optimum asphalt content by weight of total mixture, which was around 5 percent, was determined as per TxDOT procedure. This value was used in the study because it represents the

Table 1—Particle Size Distribution of Aggregates Used in This Study

U.S. Standard Sieve	Percent by Weight
Passing 12.5 mm, retained on 9.5 mm	9
Passing 9.5 mm, retained on No. 4	31
Passing No. 4, retained on No. 10	19.1
Passing No. 10 sieve	40.9

amount of asphalt cement used in the field. Three target VTM's (3.5, 5, and 6.5 percent) were chosen. Different VTM's were achieved by varying the compactive efforts in the molding process.

The variations in VTM from the target VTM for the specimens were limited to 0.3 percent. For clarity, the nominal VTM's of the specimens were used in the presentation.

The extended oven aging method was utilized to age the specimens for aging periods of 1, 2, 4, 7, 14, 21, and 28 days. Four oven temperatures (25°C, 50°C, 85°C and 100°C) were selected. The temperature of 25°C was used to verify that the specimens age insignificantly at room temperature and to demonstrate that the results are stable and repeatable.

At each oven temperature, nine pairs of specimens (with asphalt contents of 4, 5, and 6 percent and VTM's of 3.5, 5, and 6.5 percent) were placed in the oven. At the end of each aging period, the specimen to be tested was removed and quickly placed in the specimen holder so as to obtain a reading at the highest possible temperature. The travel times were recorded at 2.5°C intervals until the specimen cooled to a temperature of 25°C. The tests were repeated for every specimen for all aging periods.

Due to the nondestructive nature of the testing procedure, it was possible to test a single specimen throughout the entire aging period for one aging temperature. Changes in the elastic modulus were measured during long-term aging and the cooling process. By measuring the changes during the cooling process, a relationship between temperature and the stiffness of the HMA could be established.

RELATIONS BETWEEN MODULUS AND SPECIMEN TEMPERATURE

The stiffness of a mixture is considerably influenced by temperature. The film of asphalt cement surrounding the aggregate particles of a mixture serves as a binder or cementing agent. The asphalt cement changes from a solid to a viscous liquid as the temperature changes from room temperature to a relatively higher temperature, due to its viscoelastic characteristic. As the nature of the film changes with increasing temperature, the aggregates become less tightly bound and the mixture less rigid. For engineering purposes, temperatures at or below 60°C are of interest.

Figure 2a shows the variations in elastic modulus with temperature for a specimen at different ages. The slopes of the modulus-temperature curves are almost the same, regardless of the aging period.

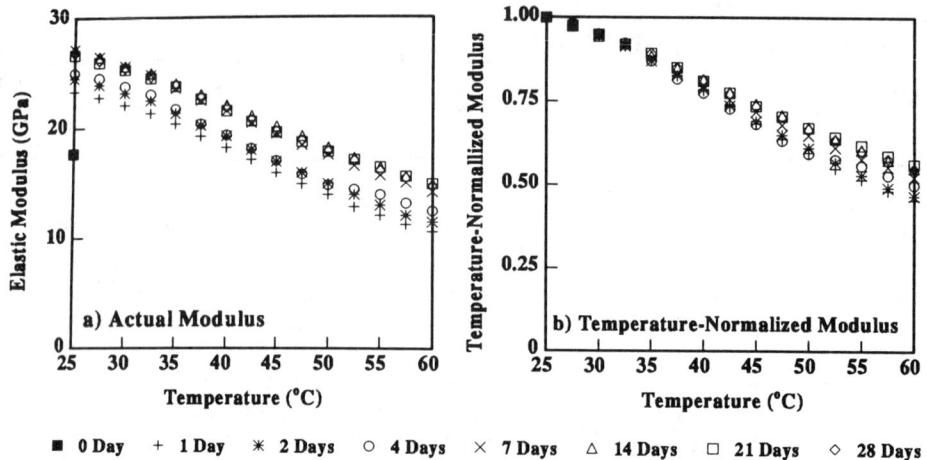

FIG. 2--Relationship between Modulus and Temperature

To determine the effects of temperature on stiffness, the "temperature-normalized modulus" (T.N.M.) was calculated using the following formula:

$$T.N.M. = \frac{Modulus\ at\ Temperature\ t}{Modulus\ at\ Temperature\ of\ 25°C} \quad . \tag{2}$$

With the introduction of the "temperature-normalized modulus," the variations in modulus with temperature at different ages can be directly compared. Figure 2b shows this variation. Data points representing different ages practically merge into a single curve. If this curve is utilized, the modulus of the mixture at different temperatures can be easily converted into a common temperature without considering the aging period.

Figure 3 shows the variations in temperature-normalized modulus with temperature for all the specimens used in this study for all aging periods (a total of 54 specimens). A fitted curve with the envelopes of a 95 percent confidence level is also shown. It is representative of a first-degree linear equation having a R^2 value of 0.89 and a standard error of 0.05. The deviation of the data from the fitted curve lessens as the temperature decreases. The maximum possible error is about 35 percent at 60°C, less than 30 percent at 50°C, and only 15 percent at 40°C.

For practical purposes, a unique first-degree linear equation can be utilized to describe the relationship between temperature and the temperature-normalized

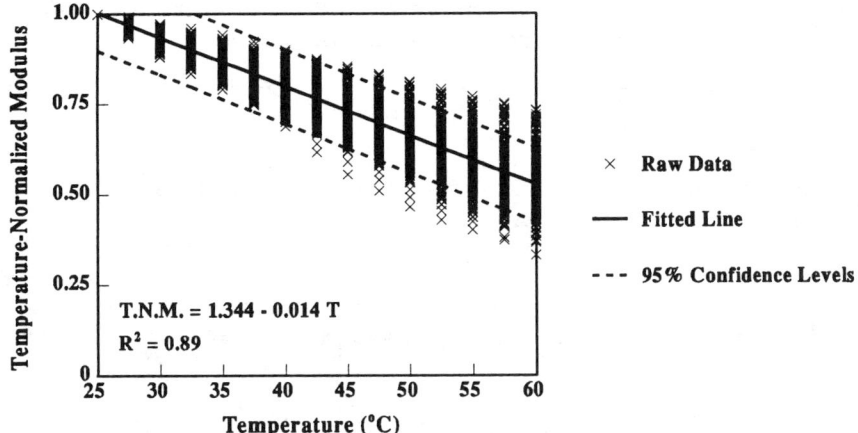

FIG. 3--Variation in Temperature-Normalized Modulus with Temperature

modulus of a mixture, irrespective of the asphalt content and VTM in the mixture. Although the slope does change with different asphalt contents and VTM's, in a practical point of view, we can still assume that a unique curve exists for a mixture at low temperatures (not to exceed 40°C). However, a more elaborate relationship can be developed, as shown below.

Figure 4a shows the variations in least-squares best-fit relationships between the temperature and temperature-normalized modulus for the specimens aged for different periods. There are seven lines in the graph. Each line represents all the specimens for one particular aging period. The R^2 values vary from 0.91 to 0.94 in all cases, which is quite reasonable.

Figure 4b shows the variations in the slope of fitted lines with aging time. The slope represents the influence of temperature on the modulus of the specimen for every aging period. As the aging period increases, the slope of the modulus-temperature curve decreases, indicating that as the asphalt cement becomes less active, the mixture becomes less influenced by temperature.

An analysis was performed to determine the correlation, if any, between different variables and the elastic modulus. The results of this analysis are presented in Table 2. As can be seen, the specimen temperature has a significant effect on elastic modulus, whereas the asphalt content and VTM have moderate effects. Finally, the aging temperature and period have only small influences on the modulus.

For the temperature-normalized modulus, as reflected in Table 2, the dominant factor is the specimen temperature. The asphalt content has a small influence, while the other factors have only negligable effects.

RELATIONS BETWEEN MODULUS AND AGING PERIOD

As mentioned earlier, the modulus of the asphalt aggregate mixture is directly related to the stiffness of the mixture. It is expected that the modulus of the mixture should increase as the aging period increases. To exhibit the variation in modulus with aging, the term "modulus ratio" is used. It is defined as

$$\text{Modulus Ratio} = \frac{\text{Modulus at t Days of Aging}}{\text{Modulus before Aging}}, \quad (3)$$

where both moduli on the right side of the equation are at a temperature of 25°C.

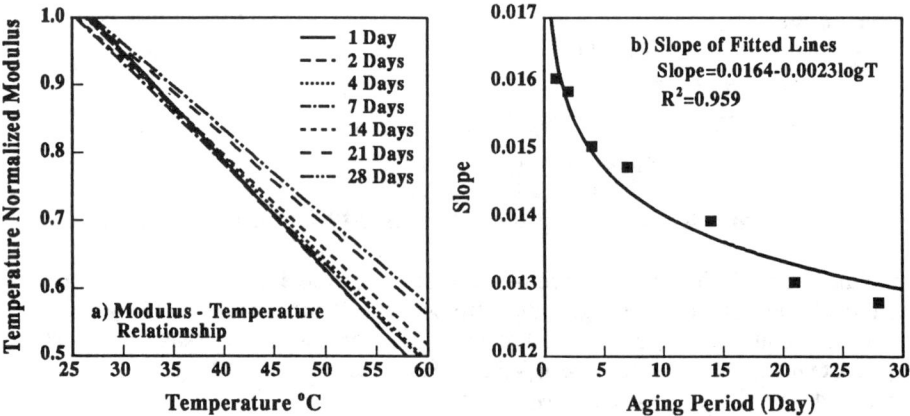

FIG. 4--Influence of Aging on Variation in Modulus-Temperature Relationship

In Figure 2, the typical variation in elastic modulus with temperature is shown for one specimen (asphalt content of 5 percent, VTM of 5 percent, and aging temperature of 85°C). As the aging period increases, the moduli of the specimens

Table 2--Correlation Coefficient Among Variables Considered in Modulus-Temperature Relationship

Parameter	Coefficient of Correlation				
	Asphalt Content	Voids in Total Mix	Aging Temperature	Specimen Temperature	Aging Period
Elastic Modulus	-0.41	-0.29	+0.19	-0.69	+0.17
Temperature-Normalized Modulus	-0.16	-0.07	-0.06	-0.95	+0.07

increase at all temperatures. The effects of temperature were discussed in the previous section. In this section, the concentration will be on the variation of modulus with aging at a constant temperature of 25°C.

To better quantify the effects of the aging period on modulus, elastic moduli at a temperature of 25°C for different aging periods were transformed into the modulus ratio using the above equation. Figure 5a shows the variation in modulus ratio with aging time for the specimen presented in Figure 2. Also shown in the figure are the results from a replicate test on a very similar specimen. Typically, the results from the tests on the two replicate specimens are quite similar and are within 5 percent of one another. The reason for the difference in moduli between the two specimens (except experimental error) after 7 days of aging is not known. Also from the figure (as well as from Figure 2), most of the increase in modulus with the aging period occurs during the first few days.

On a semi-logarithmic scale (as in Figure 5b), the modulus ratio-temperature relationship can be represented with a straight line. The slope of the fitted line depicts the rate of increase in the modulus ratio with aging. In the following sections, the effects of aging temperature, asphalt content, and VTM on aging are discussed utilizing this concept.

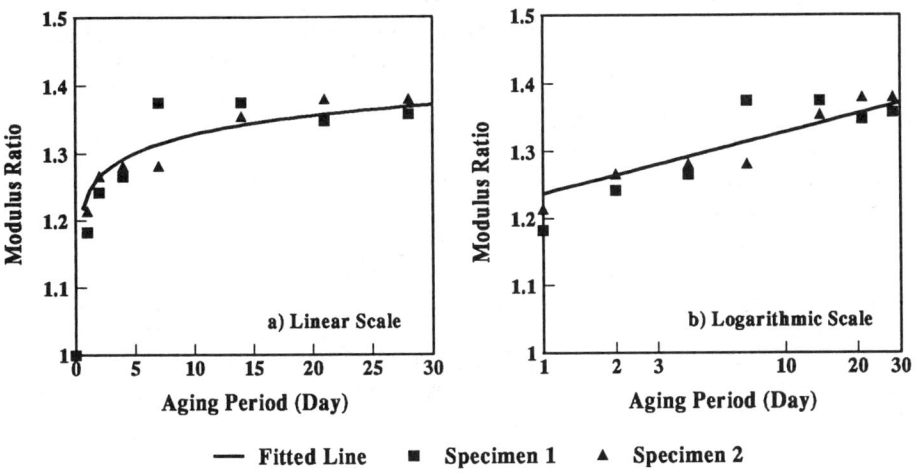

FIG. 5--Variation in Modulus Ratio with Aging Period

Effects of Aging Temperature

Figure 6 shows the variations in the modulus ratio with aging period for all specimens for an aging temperature of 85°C. Even though a general trend of increase in modulus with aging can be seen for each aging temperature, the variations in modulus for each aging period are too large to ignore the effects of other parameters.

Although not shown, the same arguments are applicable to the results obtained for aging temperatures of 50°C and 100°C.

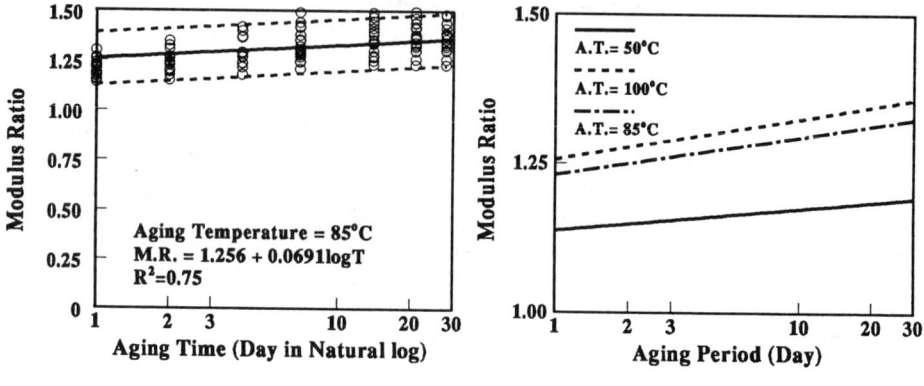

FIG. 6--Typical Variation in Modulus Ratio with Aging Period

FIG. 7--Variation in Modulus Ratio with Aging Period and Temperature

The fitted lines for the three aging temperatures are compared in Figure 7, which can be considered a rough, first-approximation relationship between modulus and aging time. As the aging temperature increases from 50°C to 85°C, the slope increases considerably; as the aging temperature increases from 85°C to 100°C, the slope does not appreciably change. This can be interpreted to mean that the aging temperatures of 85°C and 100°C produce similar degrees of aging for a mix where all other parameters are ignored. This effect is much more pronounced in this case than in the one where the aging temperature of 50°C is used.

Based upon studies conducted by other researchers (such as [13]), 50°C may not be representative of field conditions; that is, higher temperatures are required. Meanwhile, an aging temperature of 100°C may not be feasible either because of the difficulties involved in handling specimens. At such a temperature, some specimens were so soft that they could not maintain their original shape. One specimen even fell apart on its 20th day in the oven. In the following discussion, only the 85°C aging temperature results are presented and discussed.

Effects of Asphalt Content

Figure 8 shows the variations in modulus ratio with aging time at three VTM's. Each graph shows three fitted lines for different asphalt contents. As indicated earlier, the Y-intercept depicts an increase in the modulus ratio on the first day of aging, while the slope of the fitted line represents the rate of increase in the modulus ratio with aging time. In each figure, the maximum percent difference between the actual data and the fitted lines is also depicted. In almost all cases, the maximum differences are less than 10 percent. These small errors are indicative of a good fit between the data and best-fit lines.

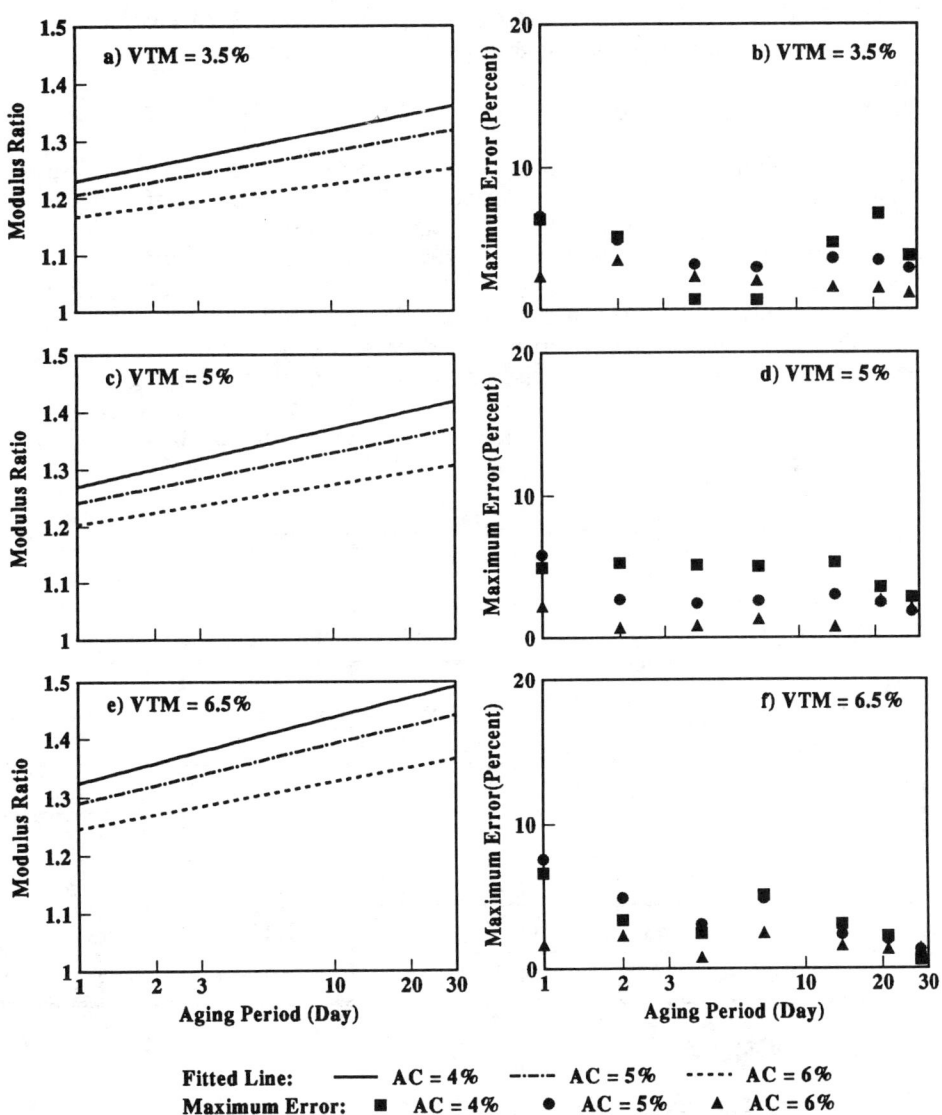

FIG. 8—Influence of Asphalt Content on Aging

Table 3 summarizes the properties of those fitted lines. At each VTM, the slope, as well as the intercept, decreases as the asphalt content increases. In other words, the aging effects are more evident for the HMA with lower asphalt contents. This result is in agreement with other studies that indicate that HMA mixtures with thicker asphalt cement films age less than mixtures with thinner films.

Effects of Voids in Total Mix

The data in the previous section are shown again in Figure 9 but are organized in a different manner. To show the influence of VTM, the specimens of different VTM's having the same asphalt content are arranged in one graph. From either Figure 9 or Table 3, it can be seen that the slope, as well as the intercept, is in direct relationship with VTM for every asphalt content. In other words, the aging effects are more obvious for the HMA with higher VTM. The results confirm that oxidization is one of the most important mechanisms of aging, because higher VTM's provide more access to oxygen, which leads to more oxidization and aging.

The intercepts of those fitted lines are in the range of 1.18 to 1.34. That means the modulus ratio increases about 18 to 35 percent during the first day of aging, depending on the VTM and asphalt content of the mixture.

Table 3--Variations in Least-Squares Best Fit Parameters with Asphalt Content and Voids in Total Mix

Nominal VTM, Percent	Fitted Line Parameters	Asphalt Content, Percent		
		4	5	6
3.5	Intercept	1.245	1.218	1.178
	Slope	0.071	0.060	0.046
5.0	Intercept	1.287	1.256	1.215
	Slope	0.078	0.069	0.055
6.5	Intercept	1.344	1.307	1.257
	Slope	0.091	0.081	0.065

Analysis of Parameters

Four variables are involved in the relationship between the modulus ratio and aging period. To determine their significance, an analysis was performed to learn of any correlation between them and the modulus ratio. The results are presented in Table 4.

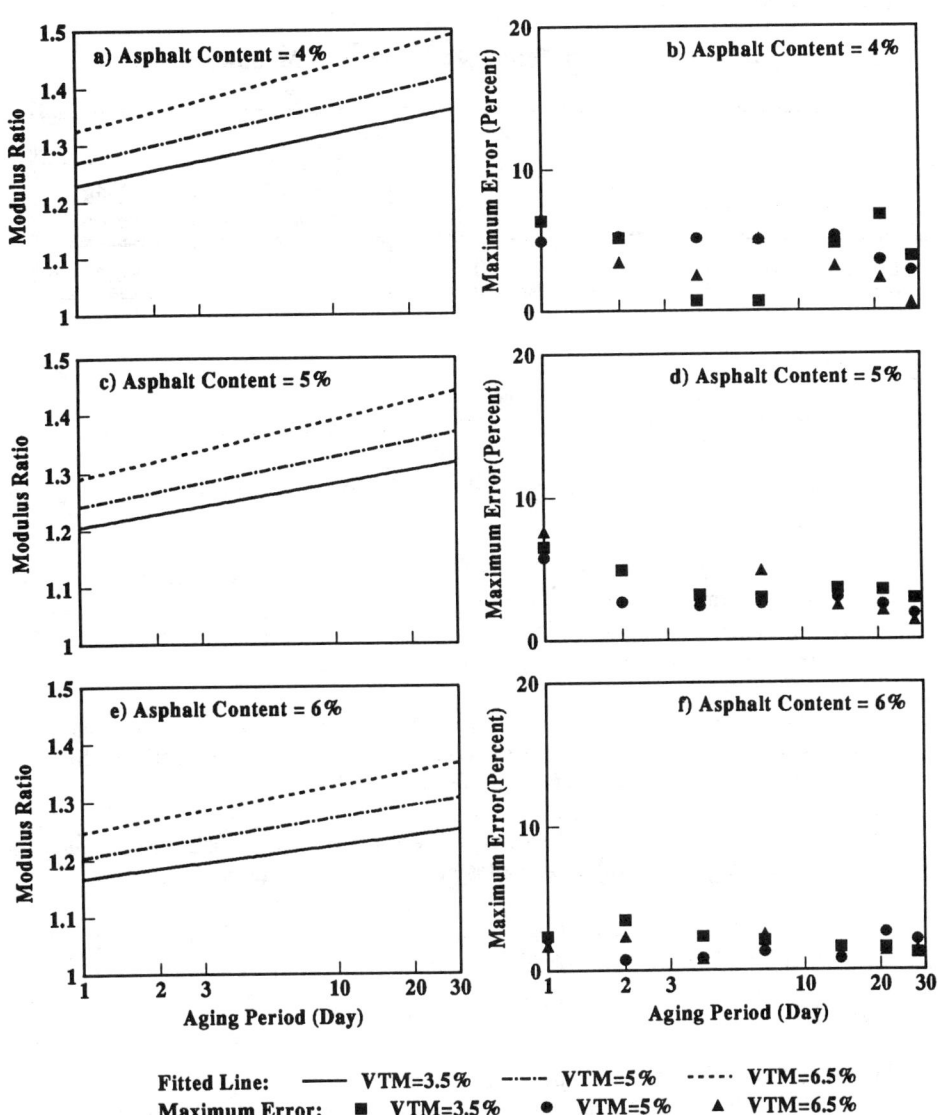

FIG. 9—Influence of Voids in Total Mix on Aging

The significance of the variables can be placed in the following order. With regards to the modulus ratio, the most important factor is aging period. Asphalt content and aging temperature have moderate effects, whereas VTM has only small effects on the modulus ratio.

Table 4--Correlation of Different Variables with Modulus Ratio

Parameter	Coefficient of Correlation			
	Asphalt Content	Voids in Total Mix	Aging Temperature	Aging Period
Modulus Ratio	-0.20	+0.08	-0.17	+0.48

CONCLUSIONS

Based on the results from this study, the following conclusions were made:

1. The wave propagation techniques can be effectively used to monitor the aging of HMA.

2. For all practical purposes, the relationship between temperature and the temperature-normalized modulus can be considered a straight line, regardless of asphalt content, VTM, or aging time. The error introduced by the fitted line decreases as the temperature decreases.

3. Based upon the correlation analysis, the specimen temperature has a dominant effect on elastic modulus. Asphalt content plays secondary role, while VTM, aging temperature, and aging period have only small effects.

4. For the temperature-normalized modulus, the dominant factor is the specimen temperature, with all the other variables having negligible effects.

5. The aging effects are directly proportional to the VTM in the HMA. They are inversely proportional to the asphalt content.

REFERENCES

[1] Petersen, J.C. " Chemical composition of Asphalt as Related to Asphalt Durability: State of Art," Transportation Research Record 999, Transportation Research Board, 1984.

[2] Kim, O., Bell C.A., Wilson J.E. and Boyle G., "Effect of Moisture and Aging on Asphalt Pavement Life," Oregon State Highway Division HP & R Study: 083: 5157, Final Report, Part 2 - Effect of Aging, 1986.

[3] Bell, C.A., AbWahab Y. and Cristi M.E., "Investigation of Laboratory Aging Procedures for Asphalt-Aggregate Mixtures," Transportation Research Record, Transportation Research Board, 1992.

[4] Bell, C.A., AbWahab Y., Kliewer J.E., Sosnovske D. and Wieder A., "Aging of Asphalt-Aggregate Mixtures," Proceedings, 7th International Conference on Asphalt Pavements, 1992.

[5] Wieder, A., Bell C.A. and Fellin M.J. , "Field Validation of Laboratory Aging Procedures for Asphalt-Aggregate Mixtures," Transportation Research Record 1384, Transportation Research Board, 1993.

[6] Harvey, J. and Monismith C.L., "Effects of Laboratory Asphalt Concrete Specimen Preparation Variables on Fatigue and Permanent Deformation Test Results Using SHRP A-003A Proposed Testing Equipment," Transportation Research Record 1384, Transportation Research Board, 1993.

[7] Kumar, A. and Goetz W.H. , "Asphalt Hardening as Affected by Film Thickness, Voids, and Permeability in Asphalt Mixtures," Proceedings, Association of Asphalt Paving Technologies, Volume 46, 1977.

[8] Tia, M., Ruth B.E. , Charai C.T., Shiau J.M., Richardson D. and Williams J., "Investigation of Original and In-Service Asphalt Properties for the Development of Improved Specifications — Final Phase of Testing and Analysis." Final Report, Engineering and Industrial Experiment Station, University of Florida, Gainesville, FL, 1988.

[9] Goodrich, J.L., "Asphalt and Polymer Modified Asphalt Properties Related to the Performance of Asphalt Concrete Mixtures," Proceedings, Association of Asphalt Paving Technologies, Volume 57, 1988.

[10] Nazarian, S., Baker M.R. and Boyd R.C., "Determination of Pavement Aging by High-Frequency Body and Surface Waves," Proceedings, 7th International conference on Asphalt Pavements, 1992.

[11] Nazarian S., Baker M.R. and Crain K., "Development and Testing of a Seismic Pavement Analyzer," Report No. SHRP-H-375, Strategic Highway Research Program, Washington, DC, 1993.

[12] Li Y., "Evaluation of Aging of Asphaltic-Concrete Using Wave Propagation Techniques," M.S. Thesis, The University of Texas at El Paso, 1994.

[13] Bell, C.A., "Summary Report on Aging of Asphalt-Aggregate Systems," Oregon State University, Report No. SR-OSU-A-003A-89-2, Prepared for the Strategic Highway Research Program(SHRP), 1990.

Robert P. Chapuis[1] and Antonio Gatien[1]

TEMPERATURE DEPENDENT TENSILE STRENGTH OF ASPHALT MIXTURES IN RELATION TO FIELD CRACKING DATA

REFERENCE: Chapuis, R. P. and Gatien, A., "**Temperature Dependent Tensile Strength of Asphalt Mixtures in Relation to Field Cracking Data,**" Engineering Properties of Asphalt Mixtures and the Relationship to their Performance, ASTM STP 1265, Gerald A. Huber and Dale S. Decker, Eds., American Society for Testing and Materials, Philadelphia, 1995.

ABSTRACT: The relative performance of conventional and polymer modified asphalt mixtures was evaluated in the laboratory and in the field. Homogeneous beams of several mixtures were tested in four-point-flexion tests at controlled temperature. The test results enabled a fast and accurate determination of how the tensile strength and apparent modulus decrease when the temperature increases. The three polymer modified mixtures had significantly higher tensile strength and modulus than the conventional mixture at any temperature. A method was developed for quantitative evaluation of the field cracking degree and its evolution with time (position, length, orientation, evolution of damage condition). Surveys were performed on several rigid base and flexible pavements of various ages. Usually, new overlays with polymer bitumen were less fissured (by a factor of 2 after three years) than overlays with conventional bitumen. It was also observed that flexible pavements with polymer bitumen show practically no cracks after three years of service. These results are encouraging for the polymer bitumen, but three years is too short for a rigorous comparison of performance.

KEYWORDS: asphalt mixture, tensile strength, cracking, temperature

The last two decades have seen a growth in traffic volumes and loads, and also many new developments in pavement construction technology. This has created needs for testing new materials and technologies in the laboratory and in the field, to document the improvement in performance and substantiate the cost effectiveness. Many characteristics of pavements may be used to evaluate various aspects of their performance. They include skid resistance, surface roughness, surface cracking, surface deformation, surface ravelling, subsurface problems, noise generation, etc. [1, 2, 3].

---

[1]Professor, Department of Mineral Engineering, École Polytechnique, P.O.Box 6079, Sta. Centre-ville, Montreal (Quebec), Canada, H3C 3A7.

[2]Professional Technician, Department of Mineral Engineering, École Polytechnique, P.O. Box 6079, Sta. Centre-ville, Montreal (Quebec), Canada, H3C 3A7.

This paper presents a laboratory and field investigation of relative performance of conventional and polymer modified asphalt mixtures. Its purpose is limited to a presentation of two points :
(1) a simple testing method to determine the tensile strength and modulus of an asphalt mixture versus temperature,
(2) a quantitative method for evaluating field cracking of pavements.

BACKGROUND

Asphalt mixtures tend to crack at some stage in their life under the combined action of mechanical stresses (due to traffic loads) and thermal stresses. It is usually agreed that measures of cracking are not easy to perform and interpret. There is no widely accepted measuring method for cracking in pavements, nor is cracking measurable by automated instruments [1]. Several evaluation techniques have been proposed based on results in Kenya [4], Texas [5, 6], Canada [7], France [2], Ontario [3], USA [8], Québec [9], Japan [10], Netherlands [11].

In this paper, cracking due to fatigue [12, 13, 14, 15, 16] is not considered. Only cracking in tension was investigated. Two aspects relative to the asphalt mixtures are especially important in this type of study: the tensile properties of the mixture that affect the induced strains, and the differences in rigidity between the pavements.

LABORATORY TESTS

Tested asphalt mixtures

Four asphalt mixtures were tested in flexion. The City of Montreal provided samples of three mixtures: (a) the polymer modified mixture of the Gilles Villeneuve race track (PISTE GV in graphs), (b) the polymer modified mixture that was used to repair the race track (PISTE REP graphs), and (c) the polymer modified mixture of the Tricentenaire boulevard (TRICEN. in graphs). Transport Quebec provided samples of one conventional mixture (no polymer), that is identified as St-Eustache (ST-EUST. in graphs).

Temperature controlled flexion tests

Test description--A controlled temperature test method was developed starting with the ASTM Standard D 1635-87 for cement concrete. Several modifications were made in order to meet the special objectives of the general program.
- the size of the tested specimen was 78 mm x 78 mm x 381 mm;
- the load was applied in ten to fifteen steps;
- each load was maintained constant during one minute;
- failure was reached in ten to fifteen minutes.

In the case of tests at 25°C and higher temperatures, the specimen was pre-heated at the test temperature in hot water during sixty minutes. Then the test was performed with the specimen in hot water at a constant test temperature up to failure. In the case of tests at temperatures lower than that of the laboratory, the specimen was first cooled (without water or ice) for five hours or more in a refrigerator or a freezing chamber at the test temperature. Then it was tested in a temperature controlled water bath for temperatures between 1 and 20°C, and in air for temperatures lower than 0°C. The sketch of the controlled temperature flexion test is given in Fig. 1.

182 ENGINEERING PROPERTIES OF ASPHALT MIXTURES

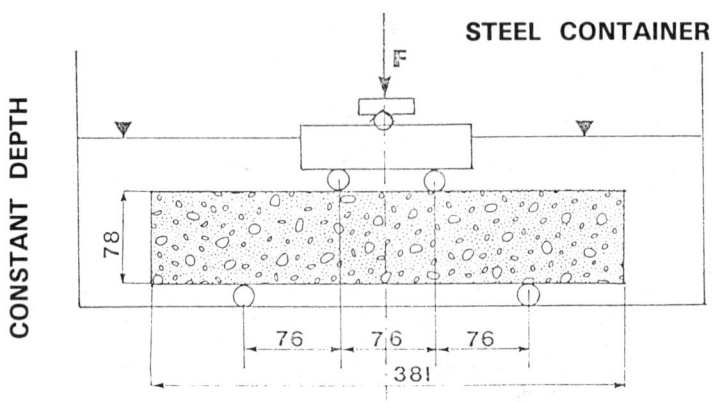

FIG. 1--Sketch of the controlled temperature flexion test (dimensions in mm).

Specimen preparation--The required masses for each specimen were prepared and oven-heated at 150°C. They were placed into a steel rigid mold and compacted in two layers, the same way as for a Marshall test. The specimen in the mold was then compacted by a test machine (200-ton capacity). The beam was prepared at exact dimensions and density within 15 to 30 seconds. It took 3 months to design, build and test the different parts of the rigid mold in order to obtain systematically homogeneous beams of asphalt mixtures.

Results for failure loads

Figure 2 gives an example of the deflexion (sag) of a beam : the mixture was of type MB-12.5, with Styrelf polymer-modified bitumen. It was compacted at 92.8 % for a relative density of 2.414. In this paper, the percentage of compaction is defined as the ratio of the bulk specific gravity (ASTM D2726) to the theoretical maximum specific gravity (ASTM D2041).

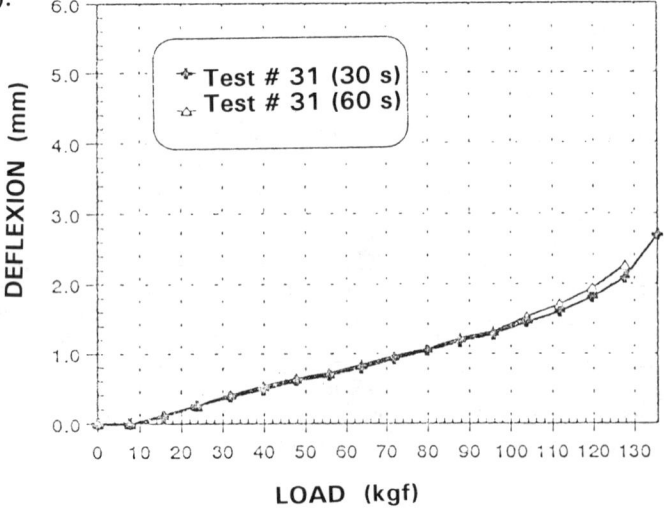

FIG. 2--Results of a temperature controlled flexion test (piste GV mixture, 25°C)

It appears that the deflexion is nearly proportional to the applied load up to 80% of the load at failure. Creep can be estimated by comparing the deflexions as measured 30 and 60 seconds after application of the constant load. Creep appears only after the applied load exceeds 70% of the load at failure. Information on creep, however, is limited, since flexion tests were not designed to evaluate creep and rheological properties of asphalt mixtures (relations between stress, strain, strain rate, temperature, aging, etc.).

In the case of the mixture for the Gilles Villeneuve race track, all loads at failure at different temperatures are given in Fig. 3, for 92 % compaction. The failure load R_{ult} decreases rapidly when temperature increases. At temperatures higher than 20°C, R_{ult} drops to half its value for a temperature increase of 10°C. It was also noticed that this mixture failed in tension by rupture between and through the aggregates (hard limestone), whatever the temperature. Broken stones were clearly visible on fracture surfaces.

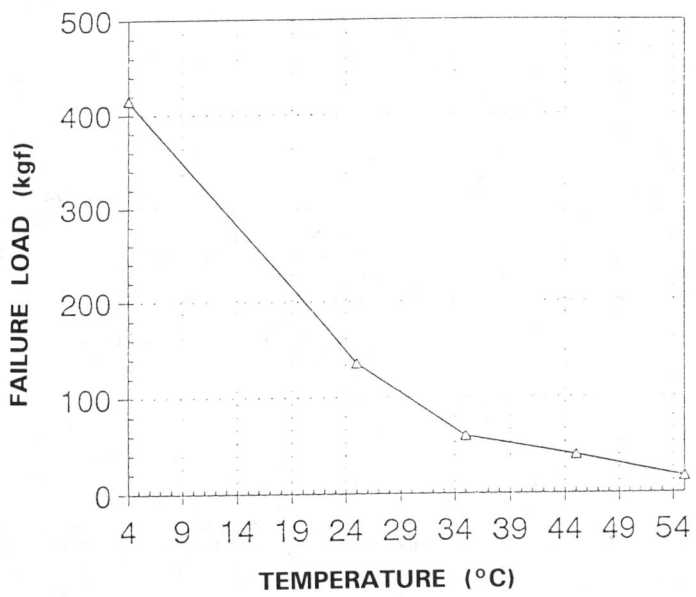

FIG. 3--Influence of temperature on tensile strength (flexion test, GV mixture).

The loads at failure for the four tested mixtures are gathered in Fig. 4. All curves have a similar shape. At temperatures higher than 20°C, each temperature increase of 10°C produces a 50% decrease in R_{ult} for the four mixtures. However, the mixture with a standard bitumen (no polymer) reached much lower values of failure load than the three other mixtures with polymer modified bitumen. The latter have similar values of failure loads.

It was impossible to do a flexion test at 55°C with the standard bitumen mixture because the beam crept and failed under its own weight. By comparison, the three mixtures with a polymer-modified bitumen had a measurable strength at 55°C, and even at 65°C (only one mixture was tested at this temperature). It is worth noting that during the summer 1991, temperatures slightly higher than 50°C were measured at the surface of pavements in the city of Montreal. Such high temperatures are critical for rutting.

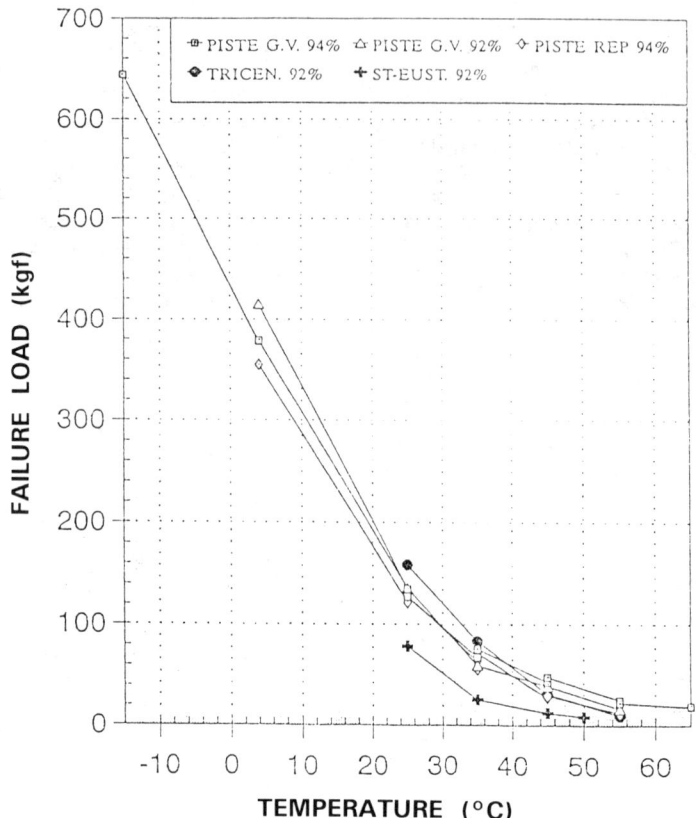

FIG. 4--Ultimate loads for flexion tests on the four tested asphalt mixtures.

In the case of the GV mixture (see PISTE G.V. 92% and 94% in Fig. 4), the density had insignificant influence on tensile strength for temperatures lower than 25°C. However, an increase in density slightly increased the tensile strength for higher temperatures.

For the three polymer-modified bitumen mixtures it was registered that the failure occurred between and through the aggregates at any temperature. Broken aggregates were clearly visible on failure surfaces. However, failure occurred only between aggregates for the standard bitumen mixture, except at 25°C for which a few aggregates were broken. This difference in failure modes, due to a difference between adhesion of bitumens to aggregates, explains the difference in failure loads.

As the number of tested mixtures was limited in this research program, many other tests would be required for a better comparative assessment of mechanical properties versus temperature of conventional and polymer-modified bitumen mixtures.

Results for moduli

The elastic modulus of a mixture beam was defined as a secant modulus between 0 and 50 % of the failure load. Figure 5 gives the moduli versus temperature for the four tested mixtures. It must be noted that these moduli are defined for quasi-static loads that are different from dynamic traffic loading conditions.

The four curves are similar: the elastic (static) modulus decreases significantly with increase in temperature. The standard bitumen mixture, however, had a much lower modulus than the three polymer-modified bitumen mixtures. The latter had comparable moduli. As previously mentioned, the testing program was limited and more tests would be necessary to draw general conclusions for modulus versus temperature.

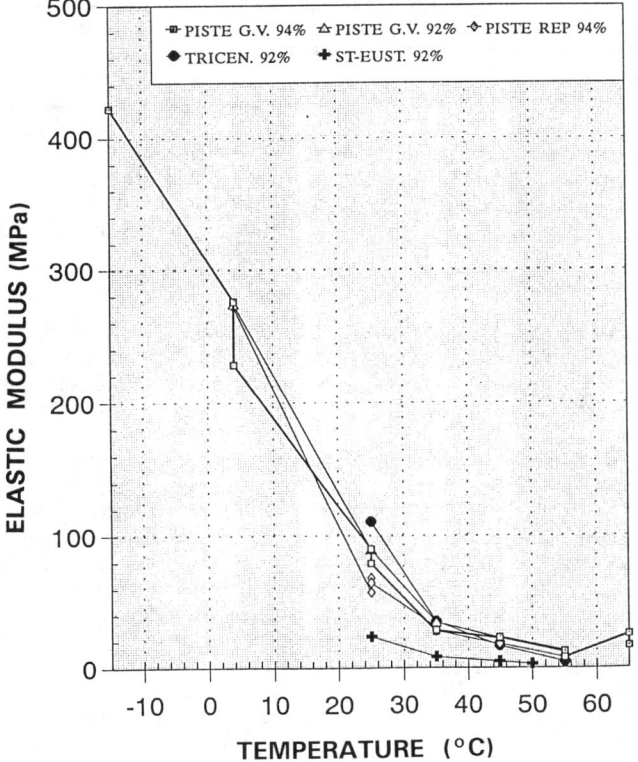

FIG. 5--Elastic modulus versus temperature.

FIELD OBSERVATIONS OF CRACKING

General considerations

Flexible and rigid base pavements (PCC pavements with asphalt mixture overlays) of various ages, were selected in the cities of Montreal and Laval (Quebec) for crack surveys (position, length, orientation, evolution of damage condition). Most cracks resulted from combined effects of thermal contraction and mixture stiffness [17, 9]. In most overlays of rigid base pavements, cracks were reflection cracks above the pre-existing cracks of the underlying (reinforced or unreinforced) concrete slab. In most cases, longitudinal and transverse cracking had already developed through shrinkage in old concrete cement slabs of rigid base pavements. For information, these slabs may be submitted to an annual temperature difference close to 60°C. The rate at which reflection cracks propagate upwards through the surfacing was of interest for the research program.

Cracking by aging effects was not found in the surveyed pavements. It is simply remembered here that aging asphalt hardens and becomes brittle. After a period of several years, strains associated with high daily temperature changes cannot be sustained and the pavement is highly fractured. Cracking by aging usually takes the form of irregular or map cracking at a spacing greater than 0.5 m, and once initiated progresses rapidly over the entire surface [13, 18]. The rate at which hardening occurs depends on several factors: the rate of oxidation of the asphalt binder, the rate at which oxygen becomes available through the voids in the mixture (air permeability), the temperature cycles and the film thickness [13]. This was not investigated in the research program.

Field data and analyzed pavements

The position of all cracks, their length and orientation were recorded, plotted on drawings, and tabulated for analysis of distribution at a given time, and analysis of evolution with time during one year. Data relative to transverse cracks were summarized for each street as follows: number of cracks per length of 100 m, mean spacing, total length, mean length. Apparent or real aperture of cracks versus temperature was not measured. Individual detailed data were presented in an internal report [19].

Most pavements in the city of Montreal include a concrete cement slab, with or without joints or dowels. Surveyed rigid base pavements had three to six lanes. A few flexible either recent or old pavements were also surveyed for a comparative analysis. Flexible pavements in Montreal are used for low volume streets and residential areas.

Recent pavements had less than three years of age during the surveys: they included rigid base pavements, mostly with polymer-modified bitumen overlays (Styrelf), and flexible pavements. Old rigid base and flexible pavements had up to twelve years of age: the bitumen of their asphalt mixtures was conventional (no polymer).

As already stated most cracks were of thermal origin, and were reflected cracks in the case of rigid base pavements. Such cracks had a neat and unique path, and an aperture of few millimetres. Very few cases of double cracks were found. Cracking by aging effects was not found. No degradation was found along the cracks.

In a few recent and hardly fissured pavements, the average transverse crack spacing was of 30 m, but more frequently it ranged between 4 and 10 m two years after construction. Statistical analysis of cracking data provided an evaluation of damage and progression of damage with time. A few results are graphically presented hereafter to illustrate the main findings.

Distribution of transersal cracks

Recent rigid base pavements--An example of distribution of transverse cracks according to their length, is given in Fig. 6 for Boulevard Pie IX, Montreal. Data are for January, June, October 1991 and January 1992. The analyzed section of 678.4 m is located between Sherbrooke and Saint-Joseph, in the southbound lanes. This is a rigid base pavement where the old cement concrete slab received a 40 mm wearing course of a mixture with polymer modified bitumen, type 13-85/100. Ten classes of transverse cracks are considered: 0-10%, 10-20%, ... up to 90-100%. The percentage represents the ratio of the crack length to the width of the pavement. It appears that the number of long cracks remained unchanged between January 1991 and January 1992. However, many new small transverse cracks appeared, and their relative length progressed between 0 and 30 % during the year of observation.

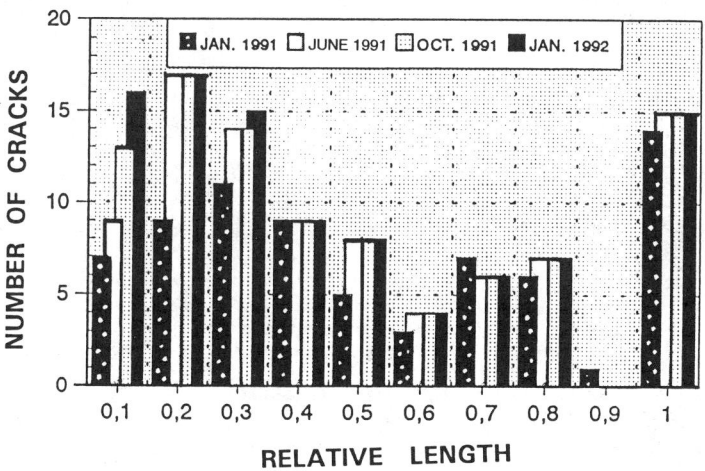

FIG. 6--Distribution of transverse cracks in Boulevard Pie IX South.

The cumulative frequency of transverse cracks is plotted in Fig. 7 for the same Boulevard. The distribution appears as relatively uniform. Such a distribution could result for example from a constant creation rate of new cracks combined with a constant propagation rate. The observation period was too short to define the creation and propagation rate functions. It is also noted in Fig. 7 that as soon as a transverse crack reaches 80 % of the street width, it crosses it immediately. Similar graphs were obtained for overlays of all recent rigid base pavements [19].

Old rigid base pavements--Figures 8 and 9 give the distribution of transverse cracks and their cumulative frequency for an old overlay (76 mm) over a rigid base pavement (6th Avenue, Montréal), 8 years after construction. The bitumen was a conventional one, without polymers. It appears in Fig. 9 that as soon as a transverse crack reaches 75 % of the street width, it crosses it completely. A somewhat different result was reached for a similar overlay (75 mm) of another old rigid base pavement (Armand Chaput St.) as shown in Fig. 10. This overlay mixture was ten years old. The cumulative frequency curve (Fig. 10) shows that, as soon as a transverse crack reaches 40 % of the street width, it crosses it completely. This indicates that the asphalt mixture of Fig. 10 is more brittle than the similar one of Fig. 9. In turn, this may indicate that cracking by aging is more likely to occur in the street of Fig. 10 than in the street of Fig. 9. The duration of surveys (one year), however, was too short to validate this interpretation.

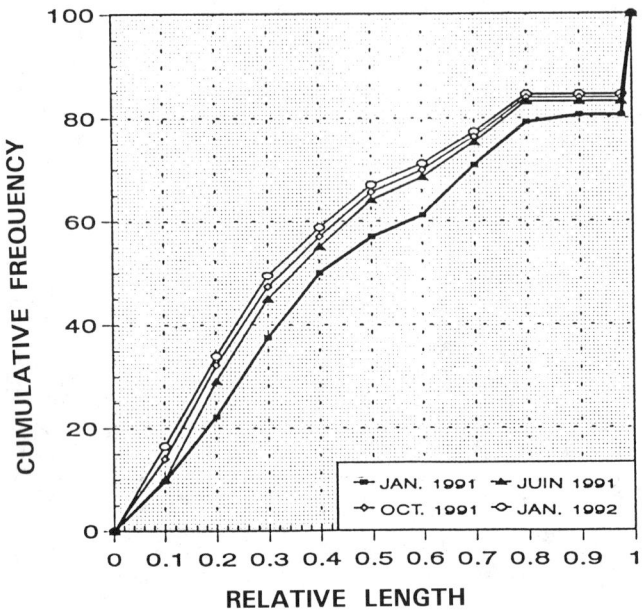

FIG. 7--Cumulative frequency of transverse cracks in Boulevard Pie IX South.

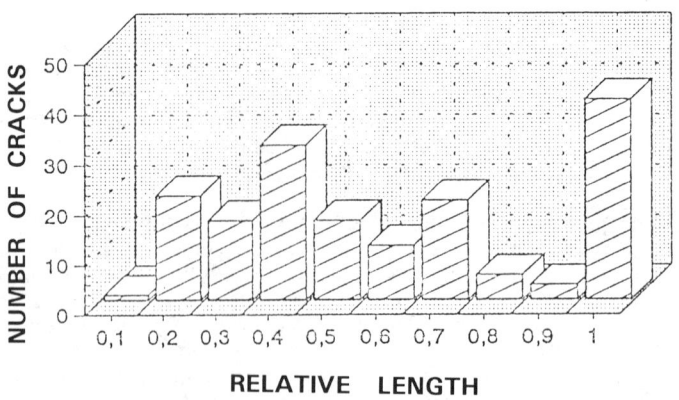

FIG. 8--Distribution of transverse cracks in the 6th Avenue (rigid, 8 years).

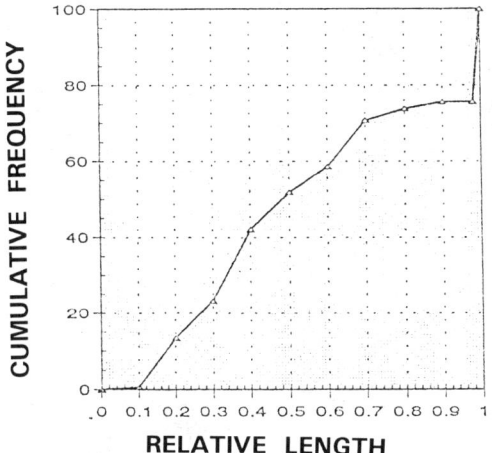

FIG. 9--Cumulative frequency of transverse cracks in the 6th Avenue.

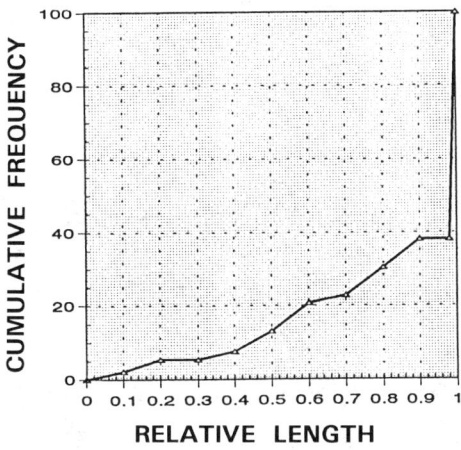

FIG. 10--Cumulative frequency of transverse cracks in a 10-year old rigid pavement.

Recent flexible pavements--They were much less fissured than overlays on rigid base pavements. Recent flexible pavements of Montreal were hardly cracked. Such a difference will be discussed below in the section on evolution of cracking with time.

Old flexible pavements--Figure 11 gives the cumulative frequency of cracks on another portion of the 6th Avenue with a ten year old flexible pavement. The asphalt mixture had no polymer.

190 ENGINEERING PROPERTIES OF ASPHALT MIXTURES

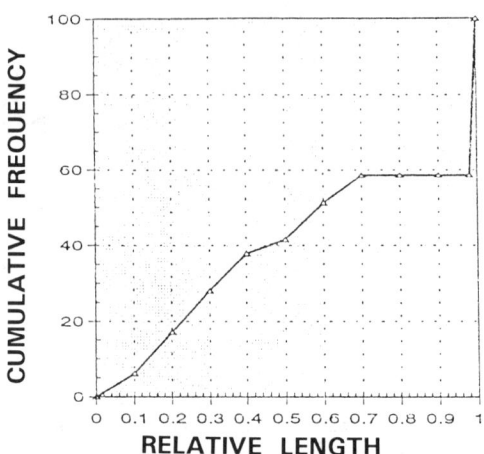

FIG. 11--Cumulative frequency of transverse cracks in a 10-year old flexible pavement.

Evolution with time of transverse cracks

For comparison between pavements, a transverse fissuration degree (TFD) was defined as the ratio of the total length of transverse cracks to the length of the pavement. Figure 12 gathers all TFDs of either old or recent pavements.

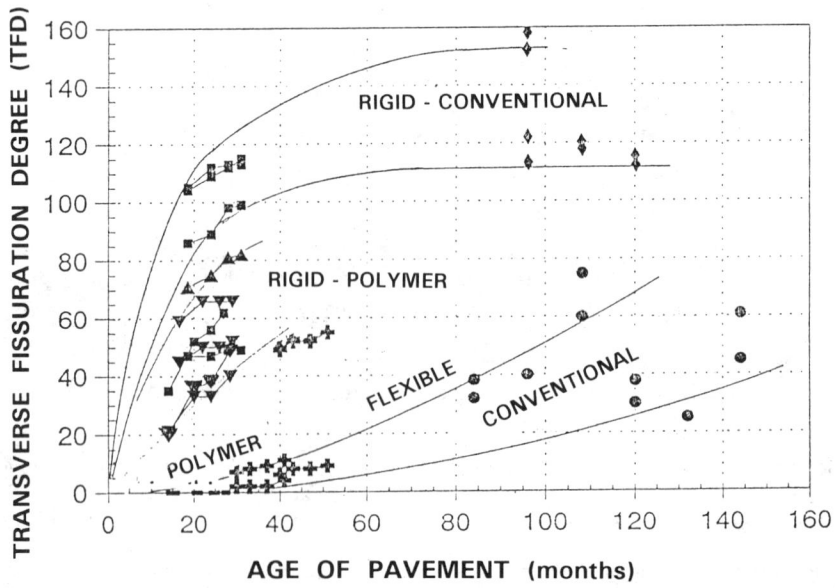

FIG. 12--Transverse fissuration degree of pavements in Montreal.

It appears that the TFDs of the two types of pavements increase with age. There is however a large difference between flexible and rigid base pavements. There are also smaller differences within each type. They may be due to the asphalt mixture thickness, or to the age and quality of the cement concrete slab.

Rigid base pavements--The overlays with a conventional bitumen are more fissured after three years than the adjacent overlays of same thickness with a polymer modified bitumen. After 2 or 3 years, a new asphalt overlay with conventional bitumen reaches a TFD that is similar to that of overlays 8 to 12-year old (Fig. 12). Some comparisons were possible for different thicknesses. For example, on Sherbrooke Street, the section with 60 mm of polymer modified bitumen mixture was more fissured than the section with 80 mm of the same mixture, as expected. In the future, it will be interesting to check whether the short-term increased resistance due to polymers can be extrapolated to long-term conditions, and whether the cracks in polymer mixtures deteriorate at a slower rate.

Flexible pavements--It can be seen in Fig. 12 that the TFD of flexible pavements is much lower than that of rigid base pavements. This TFD slowly increases during at least ten years for flexible pavements, whereas it usually reaches a maximum value after a few years for overlays over rigid base pavements.

In Montreal, four recent flexible pavements, 500 to 700 m long, with polymer modified bitumen, and mixture thickness between 53 and 109 mm, were surveyed : their TFD remained close to zero. This is excellent but cannot provide a rigourous comparison of mixtures with either polymer modified or conventional bitumen. Such a comparison was possible with two flexible pavements, over 1000 m long, surveyed in the city of Laval. The boulevard de la Concorde was surveyed at an age of 3-4 years: the westbound and eastbound lanes received 50 mm of base course plus 40 mm of wearing course, but the westbound lanes with polymer modified bitumen were three times less cracked (TFD = 2-4 %) than the eastbound lanes with conventional bitumen (TFD = 7-11 %). Two adjacent sections of boulevard Samson were also surveyed: both had 50 mm of base course plus 40 mm of wearing course. After 4-5 years, the section with polymer modified bitumen was hardly cracked (TFD = 6-9 %), whereas the section with conventional bitumen was much more cracked (TFD = 49-55 %).

DISCUSSION AND CONCLUSION

The paper describes some results of a research project jointly sponsored by Elf Canada and the Ministry of Transportation of Quebec. The objective of the research project was to evaluate the relative performance of conventional and polymer modified asphalt mixtures in the laboratory and in the field. The laboratory work included temperature controlled tensile strength tests. The field work included a detailed survey of cracks within a one year period.

A method was developed to produce homogeneous beams of asphalt mixtures. Beams of several mixtures have been tested in four-point-flexion tests at controlled temperature. The test method was specifically developed for the project. The test results enabled a fast and accurate determination of how the tensile strength and apparent modulus decrease when the temperature increases. Four mixtures were tested. It was observed that the tensile strength and the modulus were significantly higher in the three polymer modified mixtures than in the conventional mixture. This increased strength should correspond to a better resistance to cracking at low temperature and also to rutting at high temperature.

This was confirmed by the detailed surveys of cracks (position, length, orientation, evolution of damage condition) that have been performed on several rigid base and flexible pavements of various ages, including pavements built in the last three years with polymer modified bituminous mixtures. More specifically, a method was developed for quantitative evaluation of the degree of cracking and its evolution with time. New polymer bitumen overlays on rigid base pavements usually are less fissured by a factor of two after three years than the overlays with conventional bitumen. Flexible pavements with polymer bitumen are practically not cracked after three years of service. These are good results for the polymer bitumen, but three years of age is too short for comparing rigourously the performance of the two types of bitumen.

ACKNOWLEDGMENTS

The research project was funded jointly by Transport Quebec and Elf Canada. The authors thank them for their support and collaboration. They also thank the Technical Services of the City of Montreal who provided information, samples and collaboration.

REFERENCES

[1] Paterson, W.D.O. 1987. Road Deterioration and Maintenance Effects. A World Bank Publication, The Johns Hopkins University Press, Baltimore, MD.

[2] Dauzats, M., and Rampal, A., 1988. Mécanismes de fissuration de surface des couches de roulement. Bulletin de Liaison des Laboratoires des Ponts et Chaussées, Vol. 154, pp.57-72.

[3] Chong, G.J., Phang, W.A., and Wrong, G.A. 1989. Manual for condition rating of flexible pavements-distress deformations. Ontario Ministry of Transportation, Research and Development Branch, Report SP-024, Downsview, Ont.

[4] Hodges, J.W., Rolt, J. and Jones, T.E. 1975. The Kenya Road Transport Cost Study: Research on Road Deterioration. Laboratory Report 673, Transport and Road Research Laboratory, Crowthorne, England.

[5] TRDF, 1980. Final report V: Pavement and Maintenance Studies- Research on the Interrelationships between Costs of Hihgway Construction, Maintenance and Utilization. Texas Research and Development Foundation, 269 p.

[6] Lytton, R.L., Michalak, C.H. and Scullion, T. 1982. The Texas Flexible Pavement System. Proc. 5th International Conference on Structural design of Asphalt Pavements, vol.1, The Univ. of Michigan and the Delft Univ., Ann Harbor, Michigan.

[7] Anderson, K.O. 1987. Pavement Surface Condition Rating System. Roads and Transportation Association of Canada, Ottawa.

[8] SHRP, 1990. Distress Identification Manual for Long-Term Pavement Performance Studies. Strategic Highway Research Program (SHRP), National Research Council, Washington, DC.

[9] Doré, G. et Durand, J.-M. 1993. Manuel d'identification des dégradations des chaussées flexibles. Ministère des Transports, Québec.

[10] Sasaki, K., Kubo, H. and Kawamura, K. 1993. Study on reflection cracks in pavements. Proceedings of the 2nd Int. RILEM Conference, Liege, Belgium. Chapman & Hall, London, pp.237-245.

[11] Molenaar, A.A.A. 1993. Evaluation of pavement structure with emphasis on reflective cracking. Proceedings of the 2nd Int. RILEM Conference, Liege, Belgium. Chapman & Hall, London, pp..21-48.

[12] Vinson, T.S., Janoo, V.C. and Haas, R.C.G. 1989. Low Temperature and Thermal Fatigue Cracking. Report SHRP-A/IR-90-001, National Research Council, Washington, DC.

[13] Epps, J. and Monismith, C.L. 1972. Fatigue of asphalt concrete mixtures - A summary of existing information. *In* Fatigue of Compacted Bituminous Aggregate Mixtures, ASTM STP 508.

[14] Pell, P.S. 1973. Characterization of Fatigue Behavior. Special Report 140, Highway Research Board, Washington, D.C., pp.49-54.

[15] Brown, S.F., Pell, P.S. and Stock, A.F. 1977. An application of simplified fundamental design procedures for flexible pavements. Proc., 4th International Conference on the Structural Design of Asphalt Pavements, Univ. of Michigan, Ann Harbor, pp.327-341.

[16] Shell International Petroleum Company. 1978. Shell Pavement Design Manual - Asphalt Pavement and Overlays for Road Traffic. London.

[17] Andersen, P.J., Andersen, M.E. and Whiting, D. 1992. A Guide to Evaluating Thermal Effects in Concrete Pavements. Report SHRP-C/FR-92-101, National Research Council, Washington, DC.

[18] Dickinson, E.J. 1982. The performance of thin bituminous pavement surfacings in Australia. Proc. 11th Conference, Australian Road Research Board, Melbourne, pp.43-61.

[19] Chapuis, R.P. 1992. Utilisation d'un bitume-polymère pour le design, la construction et la réhabilitation des chaussées - Rapport final. Projet C.D.T. C118, 379 p., École Polytechnique de Montreal, Quebec, Canada.

Rajib Basu Mallick,[1] Randy Ahlrich,[2] and Elton R. Brown[3]

POTENTIAL OF DYNAMIC CREEP TO PREDICT RUTTING

REFERENCE: Mallick, R. B., Ahlrich, R., and Brown, E. R., **"Potential of Dynamic Creep to Predict Rutting,"** Engineering Properties of Asphalt Mixtures and the Relationship to their Performance, ASTM STP 1265, Gerald A. Huber and Dale S. Decker, Eds., American Society for Testing and Materials, Philadelphia, 1995.

ABSTRACT: Rutting is one of the predominant types of distresses observed in hot mix asphalt. Different versions of creep tests are used at present to evaluate rutting potential of hot mix asphalt mixtures. To evaluate the potential of dynamic creep to predict rutting, tests were conducted under representative conditions on laboratory samples with different levels of air voids and on field cores from pavements with known rut depths. Test were also conducted on mixes with different aggregates and aggregate gradations to identify mixes with rutting potential. When subjected to dynamic creep, samples with different air voids showed a similar response to that observed in field rutting. Good correlations were obtained between permanent creep strain and rutting rates of pavements. The dynamic creep test was able to quantify the effect of aggregate type and gradation on rutting potentials of mixes. Mixes with crushed aggregates performed better in the creep test than the mixes with uncrushed aggregates.

KEYWORDS: dynamic creep, hot mix asphalt, permanent creep strain, rutting

INTRODUCTION

Background

Rutting or permanent deformation is one of the predominant types of distresses

---

[1] Graduate Research Assistant, Auburn University, Auburn, AL 36849.

[2] Research Civil Engineer, Waterways Experiment Station, Geotechnical Laboratory, Pavement Systems Division.

[3] Director, National Center for Asphalt Technology, Auburn University, Auburn, AL 36849.

observed in Hot Mix Asphalt (HMA). Rutting occurs due to the viscoelastic nature of HMA and is caused primarily by two mechanisms - consolidation of the mix under traffic,and plastic flow of the mix. Creep testing is considered to be a basic experimental method to characterize the permanent deformation potential of HMA, since fundamental creep principles can be applied to deformation of viscoelastic mixes. At present different variations of creep tests are used to characterize the permanent deformation of HMA, and there is a need for a standard test method that is relatively simple, practical and feasible, and correlates well with field performance.

Objective

The objective of this paper is to evaluate the potential of the dynamic confined "creep" test to evaluate rutting potential of HMA.

Scope

The Phase I of this study was carried out with gravel aggregate and an AC-20 asphalt cement. Specimens with different air voids (various compactive efforts and various asphalt contents) were prepared in the laboratory and tested in the dynamic confined creep test. In Phase II field cores were obtained from pavements with measured rut depths and known traffic and tested for creep. Results were analyzed to find statistical models for predicting rutting rate.

Phase III evaluated the rutting potential of HMA mixes with different aggregate types and gradations for airfield pavements. Mix designs were performed with eleven different aggregate blends with an AC-20 asphalt cement, and samples were compacted at optimum asphalt content that produced air voids of about 4.0 percent. Samples of each of the eleven mixes were tested in the dynamic confined creep test. Results from the tests were analyzed to determine whether the test method was able to quantify the effect of aggregate type on rutting potential of the HMA mixes.

LITERATURE REVIEW

Rutting or permanent deformation of a pavement is caused by progressive movement of material under repeated traffic load through consolidation or plastic flow. Contrary to the original idea that rutting occurs primarily due to the lateral movement of the subgrade ([1]), recent studies of rutted pavements ([2], [3], [4], [5], [6]) have reported that although rutting may occur as a result of weak underlying layers, the rutting observed in existing pavements is almost entirely due to the permanent deformation in the HMA layers of the pavements. This occurrence in rutting is due to increases in truck tire pressures, axle loads, and traffic volumes ([7]).

From rutting studies it has been shown that pavements with air voids lower than 3.0 percent tend to rut while those with higher air voids do not, as long as the aggregate quality is satisfactory ([2], [5], [6], [8]) (Figure 1). Pavements with air voids significantly lower than 3.0 percent tend to rut severely. HMA pavements are constructed at approximately 7.0 to 8.0 percent air voids (9) and then further densified under traffic loads to approximate 4.0 percent air voids if the mix is properly designed.

Hot mix asphalt incorporates a blend of both elastic and viscous characteristics, and is referred to as a viscoelastic material. The behavior of such materials is dependent on time of loading and temperature (10). Under constant static or repeated

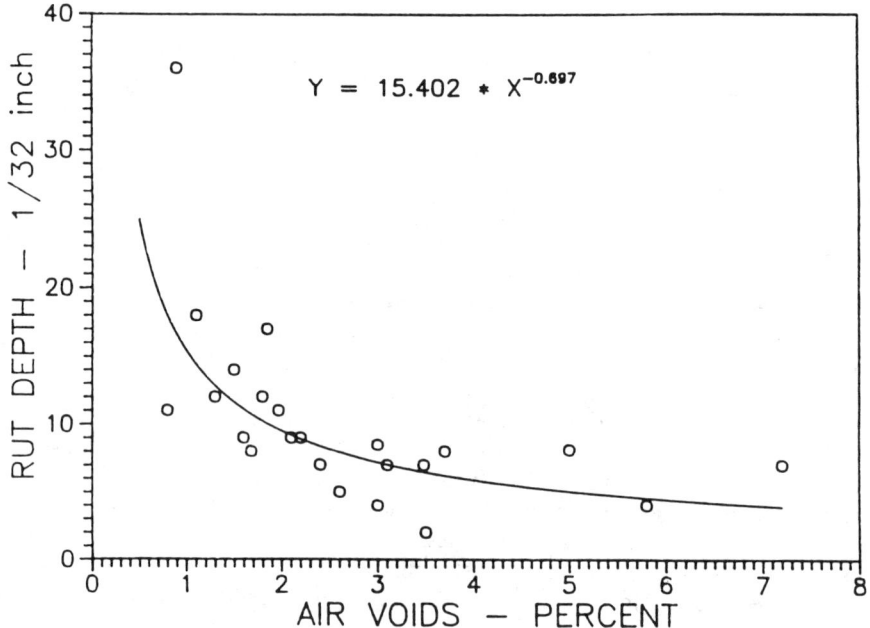

FIG. 1--HMA pavement rutting versus air voids (8)

loading, such viscoelastic material undergoes flow, or 'creep', which include recoverable and irrecoverable, time dependent and time independent components of deformation. Instantaneous elasticity, creep under constant stress, instantaneous recovery, delayed recovery, and permanent strain can be used to characterize viscoelastic materials, including HMA (11).

The principle of creep response can be applied to deformation of HMA under repeated loading. From their extensive work on creep behavior of HMA, Nair and Chang (12) stated that under repeated load in HMA there is a time dependent and time independent component of deformation. The deformation consists of three components: one recovered at removal of load, another recovered gradually and the third remaining as permanent strain. It was also reported that temperature of the HMA at the time of loading and stress level of loading significantly influence the response.

The drawbacks of assuming that viscoelastic strain is linear over the range of load applications and the time consuming process of obtaining master curves or shift factors for a particular asphalt mix can be avoided by conducting creep tests at the conditions at which rutting occurs (13).

TEST METHOD FOR DYNAMIC CONFINED CREEP TEST

A test temperature of 60°C (140°F) was used to simulate the average maximum pavement temperature throughout the United States. Truck tire pressure was simulated by maintaining a 826.8 kPa (120 psi) normal pressure and a 137.8 kPa (20 psi) confining stress. Higher deviator and confining pressures to simulate aircraft tire pressures were used for evaluating the HMA mixture with various amounts of rounded uncrushed particles. A 1378 kPa (200 psi) deviator stress along with a confining stress of 275.6 kPa (40 psi) were used to simulate aircraft traffic.

Test Equipment and Specimen Conditioning

A Material Testing System (MTS) equipment consisting of an environmental chamber, and a computerized data acquisition system, was used for the study.

For all phases of the study the specimens were maintained at the test temperature for at least two hours prior to testing. Immediately after that the specimens were preconditioned to smooth out any uneven surface and ensure good contact between the specimens and the loading plates. This was done by applying 30 cycles of 0.1 second loading and 0.9 second rest period, as recommended for the resilient modulus test (ASTM D4123). Preconditioning loads of 82.7 kPa (12 psi) were used.

The work using 826.8 kPa (120 psi) normal pressure (highway condition) was performed at the National Center for Asphalt Technology (NCAT) and the work using 1653.6 kPa (240 psi) normal pressure (airfield condition) was performed at the Waterways Experiment Station (WES). The pressure vessel used to contain the creep sample in the confined test was a triaxial cell commonly used to evaluate soils.

Test Procedure

Paper disks were placed on either side of the heated specimen, which was then placed inside the environmental chamber maintained at 60°C (140°F). The confining pressure of 826.8 kPa (120 psi) was then applied. The preconditioning load of 82.7 kPa (12 psi) was applied for 30 cycles with 0.1 second loading and 0.9 second rest period. After preconditioning the sample, the deviator stress was applied for one hour with 0.1 second load duration and 0.9 second rest period. After the one hour test the load was removed and the rebound measured for 15 minutes. The strain observed at the end of this period was reported as the permanent strain.

TEST PLAN

The objectives of Phase I of the study were to evaluate the results from dynamic confined creep tests (at 826.8 kPa, 120 psi normal pressure, and 137.8 kPa, 20 psi confining pressure), and to compare the results on laboratory specimens with expected in-place performance. A locally available gravel aggregate and an AC-20 asphalt cement were used in this study. Aggregate gradations and asphalt properties are shown in Table 1. A mix design was conducted to determine optimum asphalt content corresponding to 4.0 percent air voids at 75 blow Marshall compaction. Optimum asphalt content was determined to be 5.3 percent.

Three levels of asphalt content and two levels of compactive effort were used to evaluate mixes with different air voids. The selected asphalt contents were 1.0 percent greater than the optimum (6.3 percent), optimum asphalt content (5.3 percent), and 1.0 percent less than the optimum asphalt content (4.3 percent). Marshall specimens of 101.6 mm (4 inch) diameter were compacted with 30 and 75 blows per face and were tested for permanent deformation.

The field cores (19 cores from 8 different sites with various rutting rates) (5) were subjected to dynamic confined creep test in Phase II of the study. Test conditions were similar to those applied for testing of laboratory specimens: 137.8 kPa (20 psi) confining stress, 689 kPa (100 psi) deviator stress and 3600 load repetitions per hour.

Testing was conducted to evaluate the effect of aggregate shape and gradation on rutting potential of heavy duty airport pavement mixes in Phase III of the study. Eleven different types of aggregate blends were used in this study. The aggregate blends were fabricated in the laboratory using crushed limestone, crushed gravel, uncrushed gravel and concrete sand. The descriptions for each HMA mixture are listed in Table 2. The aggregate gradations for each mixture are listed in Table 3.

All HMA specimens for Phase III were prepared and compacted at an optimum asphalt content that produced air voids near 4.0 percent. This air void content was selected to insure that the amount of asphalt did not affect the rutting characteristics.

For Phase III the creep test was conducted with a confining pressure of 276 kPa (40 psi) and at 60 °C (140°F). A 1653.6 kPa (240 psi) axial load (1378 kPa, 200 psi deviator stress) was applied in a cyclic fashion with a 0.1 second load application and 0.9 second rest period.

Test plan for the entire study is shown in Figure 2.

TABLE 1--Gradation of aggregates and asphalt cement properties for phase I of the study.

AGGREGATE PROPERTIES	
SIEVE SIZE	PERCENT PASSING
19.0 mm (3/4 inch)	100
12.7 mm (1/2 inch)	99
9.35 mm (3/8 inch)	91
4.75 mm (No. 4)	63
2.38 mm (No. 8)	49
1.19 mm (No. 16)	36
0.59 mm (No. 30)	26
0.30 mm (No. 50)	17
0.15 mm (No. 100)	10
0.07 mm (No. 200)	5
ASPHALT CEMENT PROPERTIES	
TEST	RESULT
Absolute Viscosity at 60°C (140°F)	2083 Poise
Kinematic Viscosity at 135°C (275°F)	419 Centistoke
Penetration at 25°C (77°F)	76 (0.1 mm)
Flash Point	290°C (554°F)
Thin Film Oven Test	
% Loss	-
Absolute Viscosity at 60°C (140°F)	4431 Poise
Ductility at 25°C (77°F)	150 cm
Viscosity Ratio	-
Solubility in $C_2H_3Cl_3$	-
Specific Gravity at 15.6^0C (60°F)	1.034

TABLE 2--<u>Mix descriptions for different aggregates used in airfield mixes.</u>

MIX	DESCRIPTION
A	100 Percent Crushed Limestone
B	100 Percent Crushed Gravel
C	100 Percent Uncrushed Gravel
D	Crushed Gravel with 10 Percent Natural Sand
E	Crushed Gravel with 20 Percent Natural Sand
F	Crushed Gravel with 30 Percent Natural Sand
G	Crushed Gravel with 40 Percent Natural Sand
H	70 Percent Crushed Coarse Aggregate 100 Percent Crushed Fine Aggregate
I	50 Percent Crushed Coarse Aggregate 100 Percent Crushed Fine Aggregate
J	30 Percent Crushed Coarse Aggregate 100 Percent Crushed Fine Aggregate
K	100 Percent Uncrushed Coarse Aggregate 100 Percent Crushed Fine Aggregate

TABLE 3--Aggregate gradations used in airfield mixes.

SIEVE SIZE	FAA LIMITS	MIX A	MIX B	MIX C	MIX D	MIX E	MIX F	MIX G	MIX H	MIX I	MIX J	MIX K
19.0 mm (3/4 in)	100	100	100	100	100	100	100	100	100	100	100	100
12.7 mm (1/2 in)	79-99	90	89	89	89	89	89	89	89	89	89	87
9.35 mm (3/8 in)	68-88	77	78	78	78	77	78	78	78	78	78	77
4.75 mm (No. 4)	48-68	57	58	58	58	58	58	57	58	58	58	58
2.38 mm (No. 8)	33-53	43	42	44	42	42	44	45	42	42	42	42
1.19 mm (No. 16)	20-40	30	31	29	33	35	40	42	31	31	31	31
0.59 mm (No. 30)	14-30	23	23	26	24	29	34	36	23	23	23	23
0.30 mm (No. 50)	9-21	16	16	15	13	13	13	11	16	16	16	16
0.15 mm (No. 100)	6-16	9	11	10	9	9	8	7	11	11	11	11
0.07 mm (No. 200)	3-6	5.7	4.6	4.0	4.0	3.7	3.5	3.4	4.6	4.6	4.6	4.6

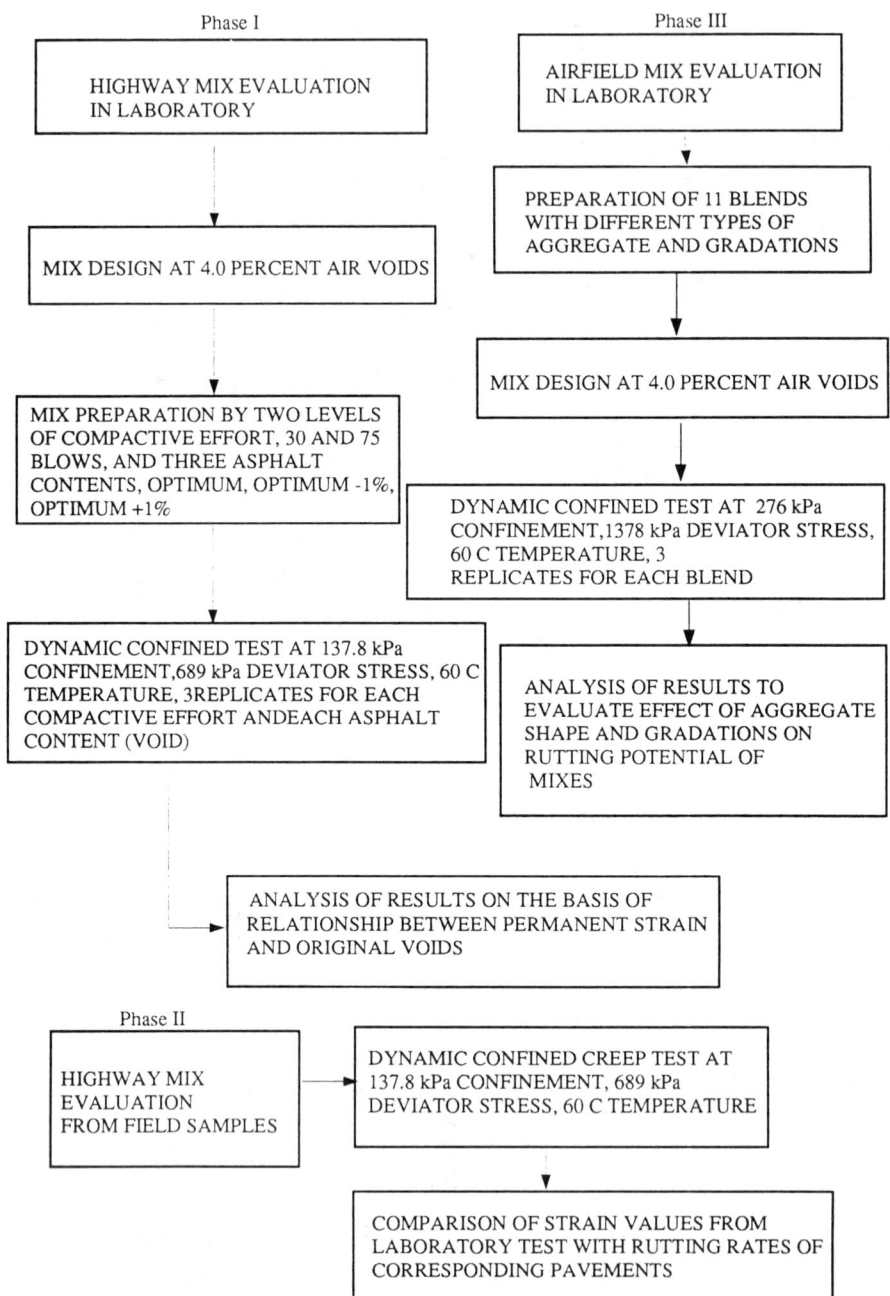

FIG. 2--Test plan.

TEST RESULTS AND DISCUSSION

Phase I of Study

Test results from dynamic confined creep tests on laboratory specimens are shown in Table 4. To observe the effect of air voids on creep due to volume change, the change in air voids during the creep test versus original air voids are plotted in Figure 3. The Y axis represents the difference between air voids of specimens before and after test. A positive value indicates an increase in voids (dilation) and a negative value indicates a decrease in voids (consolidation). Figure 3 shows that the air voids decrease when the original voids are greater than 3 percent due to low initial compaction, and increase when the original voids are less than 3 percent. This behavior is similar to that observed in the field, since it has been observed that pavements with insufficient compaction experience densification under traffic and mixtures with low voids often increase in voids when shoving takes place.
Relationship Between Permanent Creep Strain and Air voids (Voids in Total Mix, VTM)-- A good correlation of air voids and in-place rut depth is a log log relationship, as shown in Figure 1 (8). A similar best fit model, was obtained from dynamic confined creep tests of laboratory compacted samples. For both 30 and 75 blow mixes, the dynamic confined creep test results gave a good correlation with air voids (R^2 =0.94 and 0.91) as shown in Figure 4. It may be noted that pavements with air voids lower than 3.0 percent tend to rut severely (8), and this is reflected in the relationship between air voids and the results from the dynamic creep test, as shown in Figure 4.

Phase II of Study

The dynamic confined testing conditions for the field cores were selected to simulate in-situ pavement deformation. Field cores were taken outside the traffic lanes (shoulder) to evaluate the rutting potential with the creep test. Taking cores from outside the wheel path provided samples closer to the mix density when initially placed. The field cores were tested in the dynamic confined testing set-up, at 60°C (140°F), and with 137.8 kPa (20 psi) confining pressure (conditions similar to that applied for laboratory specimens in Phase I). A deviator stress of 689 kPa (100 psi) was used, with a loading time of 3600 seconds and a rebound period of 900 seconds. Each load cycle consisted of 0.1 second loading and 0.9 second unloading. The permanent creep strain values obtained from tests on cores were correlated with rut depth and rutting rate for the corresponding pavements (from which the cores were recovered). Rutting rate is defined as Rut Depth divided by square root of actual accumulated Equivalent Single Axle Load (ESAL), with the ESAL value expressed in millions. Use of rutting rate helps to normalize rutting values so that pavements subjected to various levels of traffic can be compared. Core identifications, thicknesses, permanent creep strains, and the corresponding rut depth and rutting rates are presented in Table 5.

TABLE 4--Properties of laboratory samples from phase I of the study.

SAMPLE #	AC CONTENT (%)	COMPACTIVE EFFORT (BLOWS)	INITIAL VOIDS (%)	FINAL VOIDS (%)	PERMANENT STRAIN (M/M)
1	6.3	75	1.34	3.78	0.1053
2	6.3	75	1.37	3.61	0.0731
3	6.3	75	1.46	3.32	0.0838
4	5.3	75	3.16	3.84	0.0468
5	5.3	75	2.68	3.45	0.0514
6	5.3	75	2.85	3.56	0.0665
7	4.3	75	5.86	5.73	0.0184
8	4.3	75	6.09	6.12	0.0167
9	4.3	75	5.85	5.87	0.0159
1	6.3	30	2.35	3.60	0.0897
2	6.3	30	2.03	4.33	0.1265
3	6.3	30	1.92	4.62	0.1538
4	5.3	30	5.63	4.79	0.0442
5	5.3	30	5.17	4.83	0.0457
6	5.3	30	5.09	4.93	0.0552
7	4.3	30	8.57	8.11	0.0358
8	4.3	30	8.85	8.18	0.0261
9	4.3	30	8.92	8.46	0.0393

Relation Between Rutting Rate and Permanent Creep Strain
Figure 5 shows a plot of permanent creep stain versus rutting rate of corresponding pavements. A direct relation is observed between rutting rate and laboratory strain. The best fit curves obtained through the data points are power (square) curves. Two best fit curves were plotted, as shown in Figure 5. One curve is drawn with an intercept value and the other is forced through origin. In both cases the correlations are found to be very strong (R^2 = 0.97 and 0.95). One outlier was identified and was not used in the regression analysis.

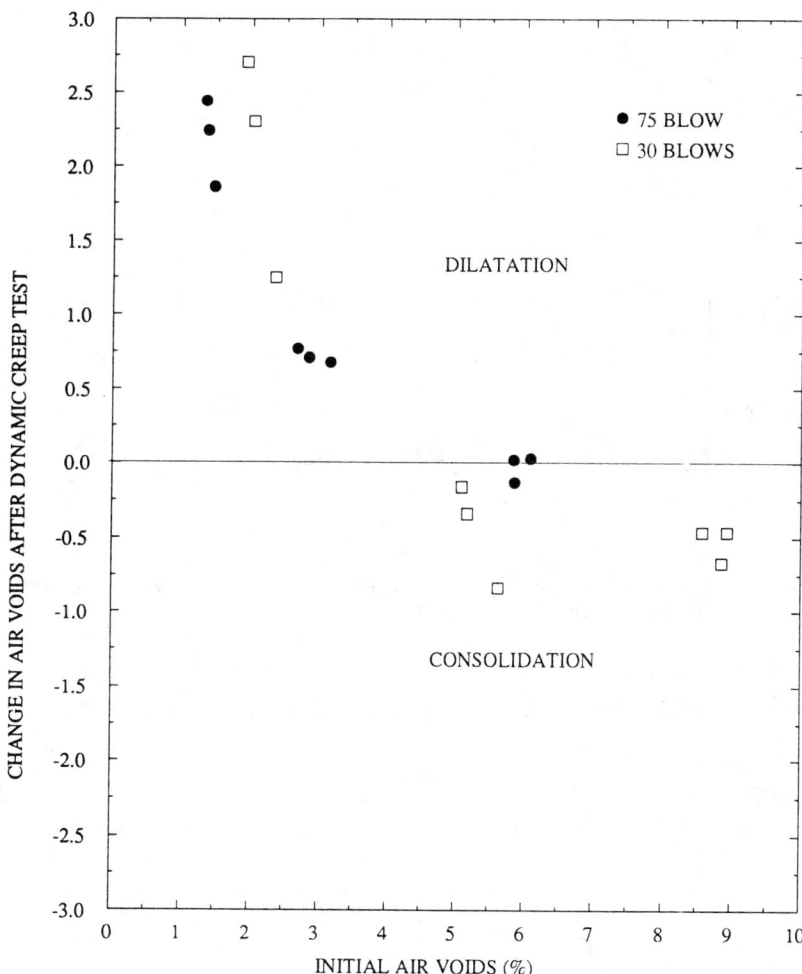

FIG. 3--Change in air voids after dynamic creep test in phase I of the study.

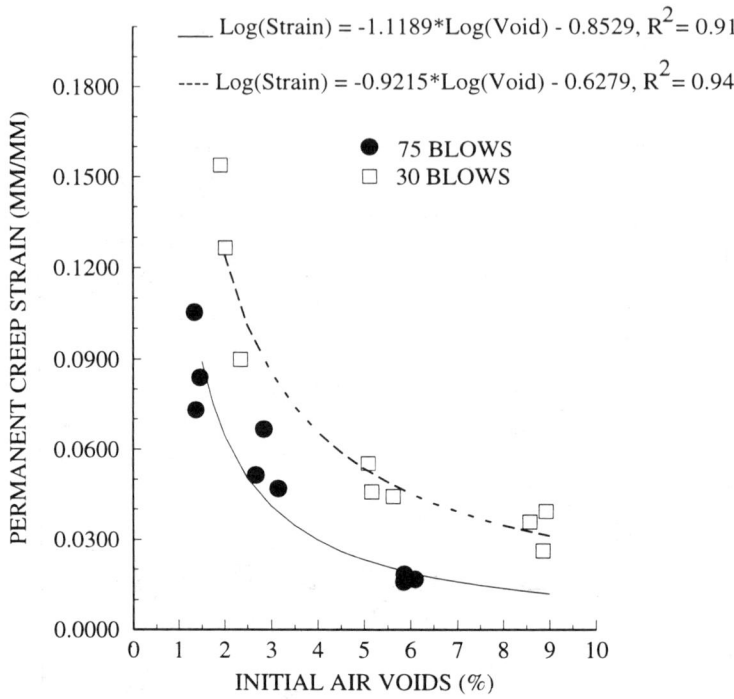

FIG. 4--Relation between initial air voids and permanent creep strain of laboratory samples in phase I of the study

TABLE 5--Field and laboratory properties of cores used in Phase II of the study

SITE	MEAN CORE THICKNESS (MM)	RUT DEPTH (MM)	RUT RATE (MM/SQUARE ROOT MILLION ESAL)	MEAN LABORATORY STRAIN (MM/MM)
10	101.19	3.17	1.94	0.02495
8	71.08	10.16	3.02	0.09117
24	74.88	8.00	3.47	0.07645
4	120.22	6.35	3.84	0.06140
23	77.09	14.88	8.19	0.13365
16	80.86	13.87	14.89	0.07625*
33	93.73	17.78	16.76	0.23380
26	81.15	13.72	No Traffic Count	0.05170

Note : * Outlier, not used in regression

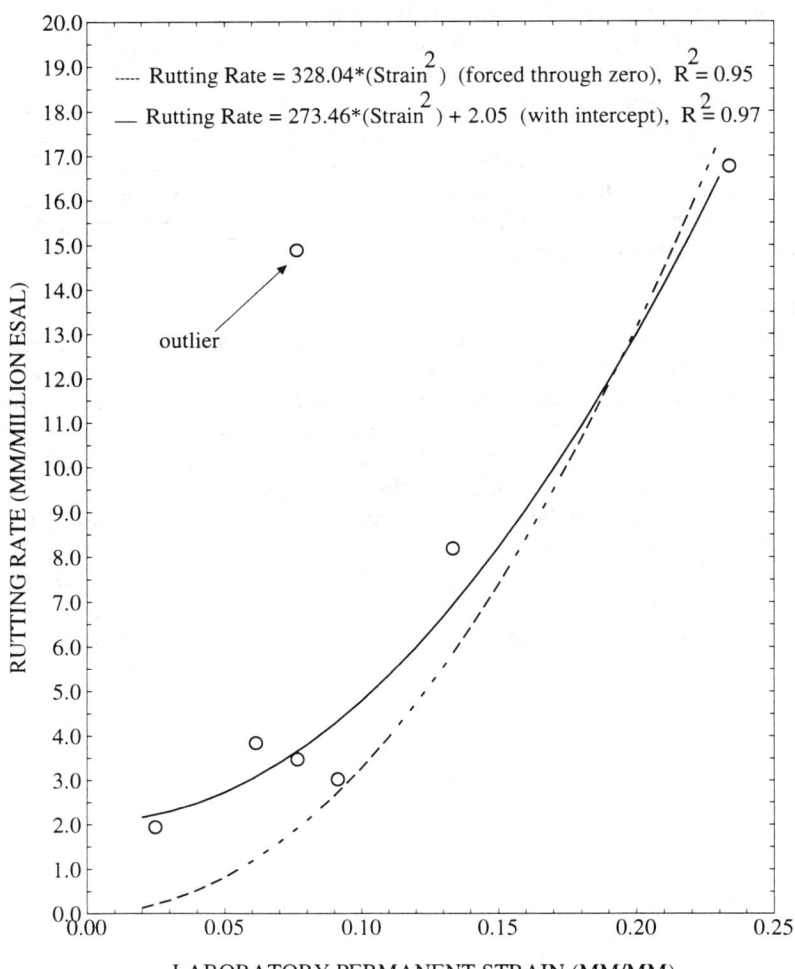

FIG. 5--Relation between rutting rate and laboratory permanent creep strain.

Phase III of Study

Particle shape of combined aggregates in an HMA mixture is a very important property which affects the in-place performance of airfield pavements. Crushed particles are desirable in HMA for improved resistance to rutting. Also, in dense graded HMA mixes fine aggregate particle shape likely contributes more to rutting resistance than coarse aggregate particle shape. Hence if any laboratory test method is to be used for predicting permanent deformation potential of HMA mixes, it should be capable of evaluating the effect of difference in aggregate shape on deformation potential. In this phase of the study, the dynamic confined creep test (1653.6 kPa, 240 psi normal pressure, and 275.6 kPa, 40 psi confining pressure) was used to evaluate mixes with different aggregate blends in terms of their rutting potential.

A summary of the dynamic confined creep test results is shown in Table 6. The permanent strain values for the different mixes are shown graphically in Figure 6. The test results indicate that the type of crushed aggregate (limestone versus gravel) and the amount of crushed particles did affect the permanent strain values. The 100 percent crushed limestone mix had less strain then the 100 percent crushed gravel mix. The strain values increased with an increase in natural sand content and a decrease in percentage of crushed coarse aggregate.

Discussion

The results provided in this paper show that the dynamic confined creep test has potential in predicting the rutting rate of HMA mixtures. This test demonstrated an ability to identify mixtures with various void contents that are known to be rutting mixes (low voids). The test also showed that it can identify mixtures, having various amounts of crushed particles, that have rutting tendencies. This test procedure also correlated with actual field performance on a number of trafficked pavements.

This test appears to be useful in the mix design of HMA mixtures and with some adaptations to simplify equipment can be useful as a field test for Quality Control/Quality Assurance (QC/QA) of hot mix asphalt (HMA) mixtures. It correlates well with field performance (based on the limited data presented) and the trends in pavement strain as a function of voids and fractured faces follow that expected for field performance.

TABLE 6--Summary of dynamic confined creep test results for airfield mixes

MIX	AIR VOIDS (%)	PERMANENT STRAIN (MM/MM)	CREEP MODULUS (KPa)
A	4.0	0.0211	78243
B	4.3	0.0352	46942
C	3.7	0.0843	19485
D	4.1	0.0350	46556
E	4.0	0.0386	42773
F	3.7	0.0384	41995
G	4.1	0.0399	41374
H	4.2	0.0407	40541
I	3.9	0.0452	36517
J	3.9	0.0455	37034
K	4.1	0.0495	33403

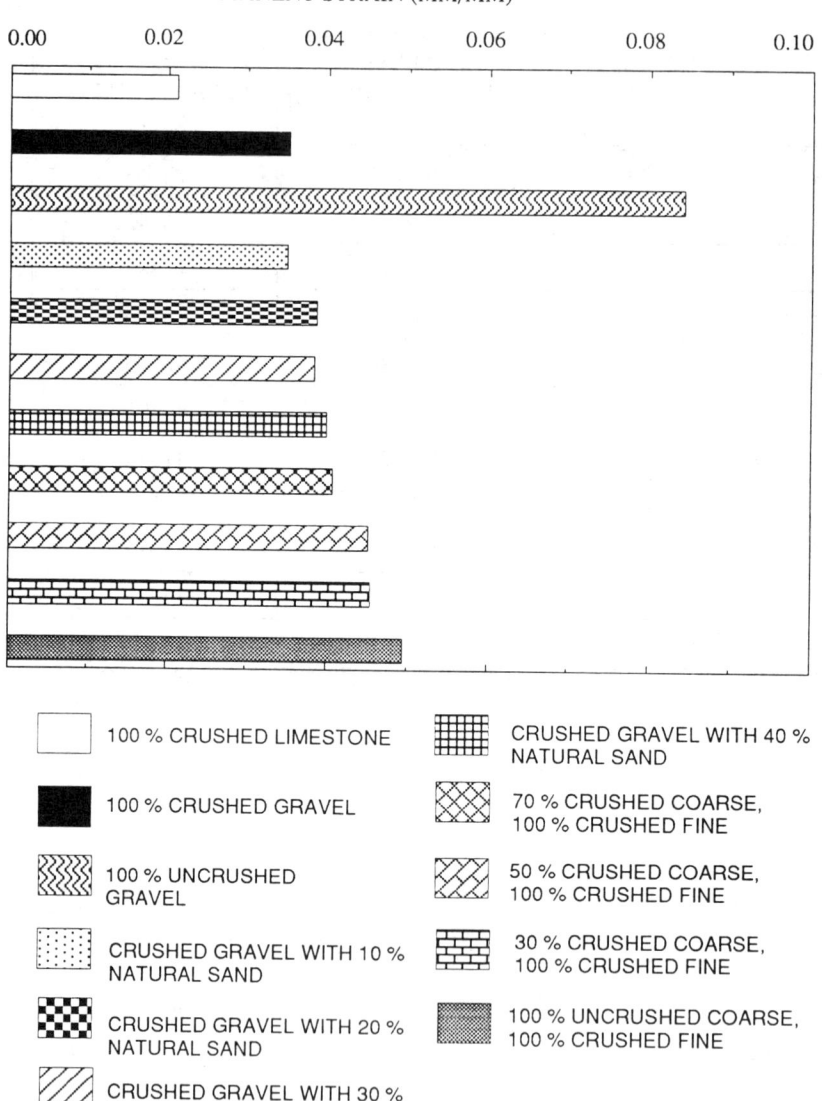

FIG. 6--Permanent creep strain values for mixes with different aggregate blends.

CONCLUSIONS AND RECOMMENDATIONS

From the results of the study the following conclusions can be made:

1. The dynamic confined creep test can simulate of field rutting since the relationship between confined creep strain and air voids of laboratory samples is similar to the relationship between in-place rutting and air voids of field cores.

2. The effect of dynamic confined creep test on laboratory samples is similar to the effect of rutting on in-place mixture. Poorly compacted samples showed a decrease in voids under the confined creep test. This is similar to what is observed in field. Samples with low voids often showed an increase in voids after testing which again match field observations.

3. Plots of strain versus air voids from dynamic confined creep tests show the existence of a critical air void level of approximately 3.0 percent, below which strain increases rapidly.

4. The dynamic confined creep test can be used to predict rutting potential of HMA mixes. Strain values from laboratory tests showed a very strong correlation with in-place rutting rate.

5. The dynamic confined creep test was used successfully in obtaining significant difference in rutting potential of mixes with different amounts of rounded aggregate. The mixture having more angular aggregates performed better as would be expected in the field. It can be concluded that the dynamic confined creep test can be used effectively as a laboratory tool to evaluate rutting potential of mixes with different aggregate blends.

Considering its effectiveness as a tool for predicting rutting, the dynamic confined creep test is recommended as a potential test to be used to evaluate rutting potential of HMA mixes in the mix design and quality control of HMA mixture. Data provided in this report indicate that this test can be used to estimate the rutting rate of HMA.

REFERENCES

(1)   The AASHO Road Test: Report 5 - Pavement Research, Highway Research Board Special Report 61E, National Research Council, Washington D.C., 1962.

(2)   Huber, G. A. and Heiman, G. H., "Effects of Asphalt Concrete Parameter on Rutting Performance: A Field Investigation", Proceedings, Association of Asphalt Paving Technologists, Volume 56, 1987.

(3)   Anani, Bassam A., Balghunaim, F. A., and Abdulrahman, S., "Laboratory and Field Study of Pavement Rutting in Saudi Arabia", Transportation Research Record,

Chip Seals, Friction Courses, and Asphalt Pavement Rutting, Transportation Research Record 1259, Transportation Research Board, Washington D.C., 1990.

(4) Lee, Kang W., and Al-Dhalaau, Mohamad A., "Rutting, Asphalt Mix Design and Proposed Test Road in Saudi Arabia", Implication of Aggregates in The Design, Construction and Performance of Flexible Pavements, STP 1016, American Society for Testing Materials, 1989.

(5) Brown, E. R., and Cross, Stephen A., "A Study of In-Place Rutting of Asphalt Pavements", Proceedings, The Association of Asphalt Paving Technologists, Volume 59, 1990.

(6) Parker, Frazier, and Brown, E. R.,, "Effects of Aggregate Properties on Flexible Pavement Rutting in Alabama", Effects of Aggregate and Mineral Fillers on Asphalt Performance, ASTM STP 1147, American Society for Testing and Materials, Philadelphia, 1992.

(7) Proceedings, A Symposium/Workshop on High Pressure Truck Tires, American Association of State Highway and Transportation Officials and Federal Highway Administration, Austin, Texas, February 1987.

(8) Ford, Miller C., "Pavement Densification Related to Asphalt Mix Characteristics", Flexible Pavement Construction, Transportation Research Record 1178, Transportation Research Board, Washington D. C., 1988.

(9) Brown, E. R., "Density of Asphalt Concrete - How Much is Needed?" Transportation Research Record 1282, Transportation Research Board, Washington D. C., 1990.

(10) Pomeroy, C. D., "Creep of Engineering Materials", A Journal of Strain Analysis Monograph, Mechanical Engineers Publication Limited, London, 1978.

(11) Findley, W. N., Lai, J. S., and Onaran, K., "Creep and Relaxation of Nonlinear Viscoelastic Materials", North Holland Publishing, Amsterdam, New York, Oxford, 1976.

(12) Nair, K., and Chang, C-Y., "Flexible Pavement Design and management Materials Characterization", Highway Research Board, NCHRP Report 140, 1973.

(13) Brown, E. R., and Foo, K, Y., "Comparison of Unconfined and Confined Creep Tests for Hot Mix Asphalts", Journal of Materials in Civil Engineering, American Association of Civil Engineers, 1994.

Koon M. Chua[1] and Myung C. Roo[2]

COMPREHENSIVE CHARACTERIZATION OF PERFORMANCE-RELATED PROPERTIES OF ASPHALT CONCRETE MIXTURES THROUGH DYNAMIC TESTING

REFERENCE: Chua, K. M. and Roo, M. C., "Comprehensive Characterization of Performance-Related Properties of Asphalt Concrete Mixtures Through Dynamic Testing," *Engineering Properties of Asphalt Mixtures and the Relationship to their Performance, ASTM STP 1265*, Gerald A. Huber and Dale S. Decker, Eds., American Society for Testing and Materials, Philadelphia, 1995.

ABSTRACT: It is often desirable to be able to obtain a comprehensive characterization of the performance-related properties of asphalt concrete with as few tests as possible. The New Mexico State Highway and Transportation Department is interested in obtaining a comprehensive characterization of the performance-related properties of the four types of asphalt concrete mixtures that are commonly used in the state. These properties include: the strength, the resilient modulus, the rutting characteristics, and the fatigue/cracking characteristics. Typically, different tests are needed to determine these characteristics. However, the approach taken here to obtain the desired information is through dynamic testing with large (15 cm diameter x 30 cm high) cylindrical asphalt concrete specimens at four different load levels, frequencies, and temperatures. The load applied were 1112 N, 2224 N, 4448 N, and 8896 N; at frequencies of 1 Hz, 4 Hz, 8 Hz, and 16 Hz. Test temperatures were 4.4°C, 25°C, 37.8°C, and 60°C. Continuous haversine load cycles were applied for each test set and the response to the repeated loadings were recorded. Resilient modulus histories were obtained. Rutting characteristics of the material at different temperatures were obtained from the residual deformation histories. The thermal viscoelastic properties were determined from the deformation response at the different temperatures. The change in the damping characteristics with repeated loading were determined through analysis of the data in the frequency domain. Since damping properties can be related to the embrittlement and aging characteristics of materials, the fatigue properties were also inferred. Additionally, it is shown that the degree of susceptibility of the asphalt concrete to cracking and reflection cracking can also be estimated.

KEYWORDS: dynamic testing, modulus, damping, cracking, rutting, fatigue

---

[1]Asst. Professor, Dept. of Civil Engineering, University of New Mexico, Albuquerque, NM 87131.

[2]Graduate Student, Dept. of Civil Engineering, University of New Mexico, Albuquerque, NM 87131.

INTRODUCTION

It would be desirable to highway engineers if a comprehensive characterization of the performance-related properties of asphalt concrete mixtures can be obtained simply with one method of testing. Even if the information obtained may not be exact, it will still enable some reasonable estimate of the performance of the mixtures to be made. The New Mexico State Highway and Transportation Department [NMSH&TD] is interested in obtaining a comprehensive characterization of the performance-related properties of the four types of asphalt concrete mixtures that are commonly used in the state. This paper presents the results obtained and also describe how it may be possible to accomplish this using only dynamic testing.

In 1986 when the American Association of State Highway and Transportation Officials [AASHTO] adopted the mechanistic pavement design procedures [1], obtaining the resilient behavior of each flexible pavement layer become necessary and dynamic testing of the asphalt material became a common method of testing.

The performance-related properties that are of interest to pavement engineers include: the strength, the resilient modulus, the rutting characteristics, and the fatigue-cracking characteristics. The strength and resilient modulus are typically determined by diametral testing of disks and sometimes at different temperatures. Fatigue is the loss of strength and other important properties at a stress level well below the ultimate strength of the material over a period time. This characteristic can manifest itself in terms of rutting and/or cracking. The fatigue response of asphalt paving mixtures can be obtained from the following tests [2]: (1) repeated flexure test, (2) direct tension test, (3) diametral repeated load test, (4) dissipated energy method, (5) fracture toughness test, (6) wheel tracking test (laboratory and field), and so forth. However, none of these tests is as yet universally accepted as a laboratory standard for fatigue characterization. The tests mentioned here are usually costly and difficult to perform on a routine basis.

The objective of this study is to determine the comprehensive performance characteristics of the different types of asphalt concrete mixtures for the state highway department and also demonstrate how these characteristics can be obtained from the dynamic modulus test. The dynamic testing was performed with large diameter cylinders (15 cm diameter x 30 cm tall) with the loading direction in the longitudinal axis. The two main reasons for selecting this test configuration are: firstly, the diameter was selected to be large enough to accommodate aggregate sizes of up to 25 mm, and secondly, the direction of load application was selected to match since the specimens were compacted vertically as in the field.

TEST MATERIAL

Four different types of asphalt concrete mixtures were obtained from NMSH&TD. Asphalt concrete mixtures of Type I-A, I-B, I-D, and ATOGB (Asphalt Treated Open Graded Base) according to the NMSH&TD Supplemental Specification Section 401 - Plant Mix Bituminous Pavement (PMBP) [3] were collected from batching plants at different work sites. Aggregate gradation for each mixture is shown in FIG. 1. The upper and lower bound for the acceptable gradation curve is also shown. Table 1 shows the design data for the different mixtures.

TABLE 1--Bituminous mix design data

Item	Type I-A	Type I-B	Type I-D	ATOGB
Asphalt Grade	AC-20	AC-10	AC-40	AC-20
Asphalt Content, %	4.1	4.5	5.0	2.07
Bulk Specific Gravity	2.377	2.374	2.344	
Unit Weight, kN/m^3	23.296	23.265	22.982	18.035
Maximum Specific Gravity	2.475	2.473	2.441	
Air Voids, %	4.0	4.0	4.0	
VMA, %	12.3	13.0	13.3	
Marshall Stability, kN	12.365	13.13	13.521	
Flow, 1/100 cm	25.4	22.9	30.5	

EXPERIMENTAL PROCEDURES

The experimental phase consists of testing of the four types of asphalt concrete. The goals of these tests are to determine the visco-elastic effects at low levels of stresses, and to determine the dynamic moduli, $|E^*|$ of the material within that stress range at different temperatures.

In the dynamic tests, unconfined cylindrical specimens of the material, 15 cm in diameter and 30 cm in height were subjected to haversine stresses of different amplitudes and frequencies, and the resulting axial strains were analyzed for their amplitudes and phase differences. This was done using an Instron Universal Testing Machine.

The test procedure for the dynamic modulus test is in accordance with ASTM (1989), "D3497-79 Standard Test Method for Dynamic Modulus of Asphalt Mixtures". In the ASTM Method, a haversine compressive stress is applied to the specimen and repeated for a minimum of 30 sec. and not more than 45 sec. at temperatures of 5°C, 25°C, and 40°C and at load frequencies 1 Hz, 4 Hz, and 16 Hz at each temperature. The axial strains are measured by bonding two wire strain gages at mid-height of the specimen opposite each other.

Table 2 shows different testing configurations used by Papazian [4], Shook and Kallas [5]. The test configuration used in this study exceeds that of ASTM (1989) by including a bigger sample size (15 cm x 30 cm) to accommodate large size aggregate (larger than 25 cm), and a higher temperature (60°C). This higher temperature in the asphalt concrete is not unusual in southern New Mexico during summer. One more frequency (at 8 Hz) is employed between 4 Hz and 16 Hz in order to obtain a more tendency of the frequency dependence and LVDTs (linear variable differential transducers) are used in place of strain gages to measure axial displacement in practical reason.

Table 3 compares the test configuration used here with the ASTM D3497-79(89). A commercially available electro-pneumatic hammer (HILTI TP800) with a circular tamping plate attached was used to prepare the test specimens. The specimens were prepared in 6 layers and each layer was vibratorily compacted for about 20 sec.

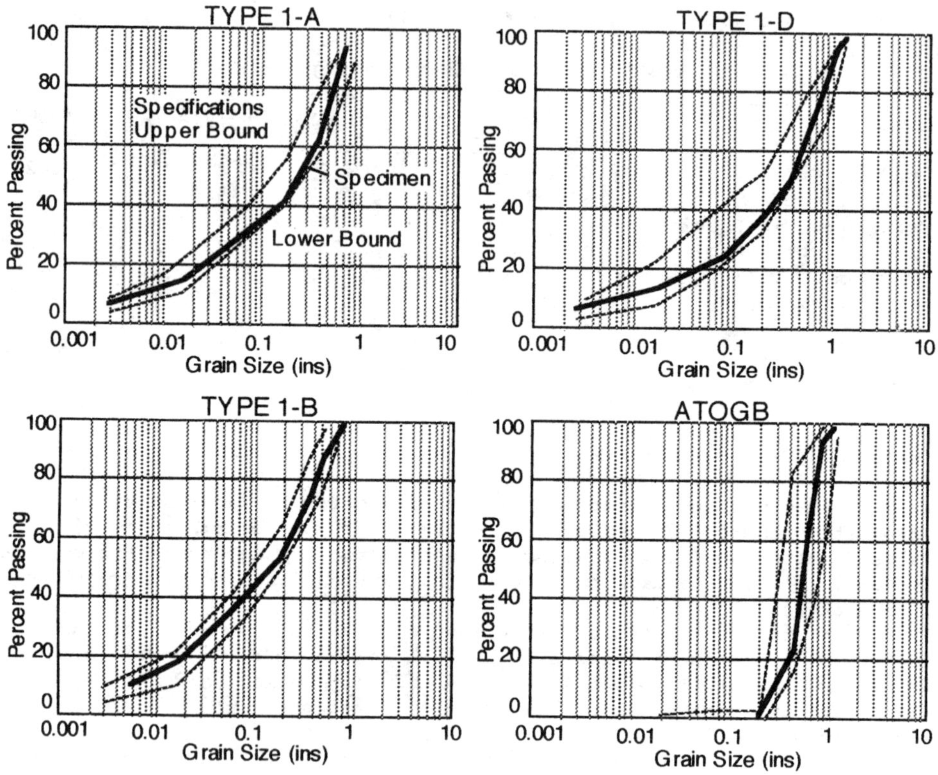

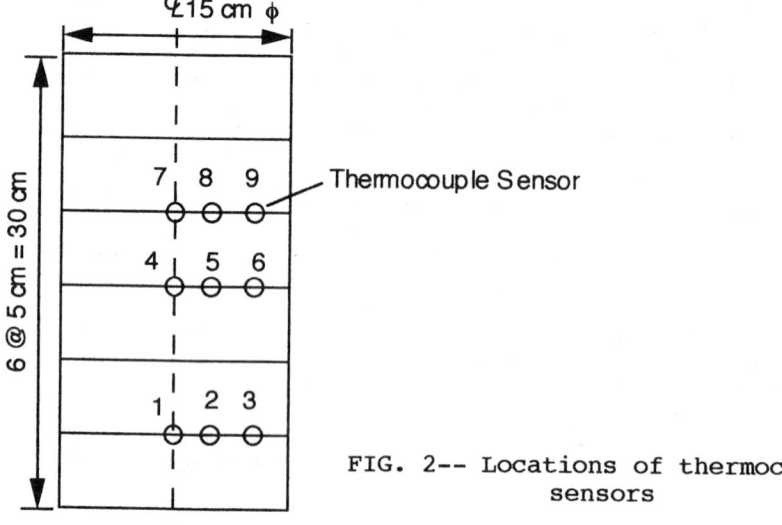

FIG. 1-- Gradations of Aggregates

FIG. 2-- Locations of thermocouple sensors

TABLE 2--Comparison between several dynamic testing methods for asphalt concrete mixture

	Year	Specimen Size (cm)	Frequency (Hz)	Temperature (°C)	Axial Stress (kPa)
ASTM	1986	10x20 (min)	1, 4, 16	5, 25, 40	0 to 241.3
Papazian	1962	7x15	1.7, 3.4, 13.7	5, 25, 40	27.6, 55.2, 110.3, 224.1
Shook & Callas	1973	10x20	1, 4, 14	4.4, 21.1, 37.8	121.4, 241.3, 482.6
Authors	1994	15x30	1, 4, 8, 16	4.4, 25, 37.8, 60	60.7, 121.3, 241.3, 482.6

TABLE 3--Differences between the ASTM and the authors' procedure

Item	ASTM D3497	Authors'
Temperature, °C	5, 25, 40	4.4, 25, 37.8, 60
Frequency, Hz	1, 4, 16	1, 4, 8, 16
Axial Stress, kPa	0 to 241.3	60.7, 121.3, 241.3, 482.6
Strain Measurement	Strain Gages	LVDTs
Capping	Sulfur Mortar	Dense Grade: No Capping (Cutting) Open Grade: Sulfur Mortar
Loading Duration	30 to 45 sec.	45 sec.
Testing Order	Temp. - low to high Freq. - high to low Load - low to high	Same Same Same

The theoretical compaction energy per layer is about 7740 kJ/m^3 [6]. The target unit weight of each mix is shown in Table 1. This is a swift and efficient way to prepare test specimens. After a 24-hour curing period at room temperature (25°C), the top portion of the each specimen is cut with the concrete saw. This was done to avoid using a sulfur mortar cap since the stiffness between asphalt concrete mixture and sulfur mortar may be large. In the case of the ATOGB, a sulfur mortar caps were required because of the loosely bound nature of the material.

Prior to testing at other than ambient temperature, the time-temperature characteristics of a typical test specimen was determined. Nine thermocouples were embedded in a Type I-B 15 x 30 cm specimen in the manner illustrated in FIG. 2. The temperature distribution in the oven or in the refrigerator, were obtained. FIG. 3a and FIG. 3b show the isotherms of a test specimen placed inside a refrigerator and a oven, respectively. It was determined that a minimum of 3 hours is required for the specimen at 25°C to reach a steady-state temperature of 4.4°C in a refrigerator and 37.8°C in a oven; and at least 4 hours to heat the specimen from 25°C to 60°C in a oven.

218 ENGINEERING PROPERTIES OF ASPHALT MIXTURES

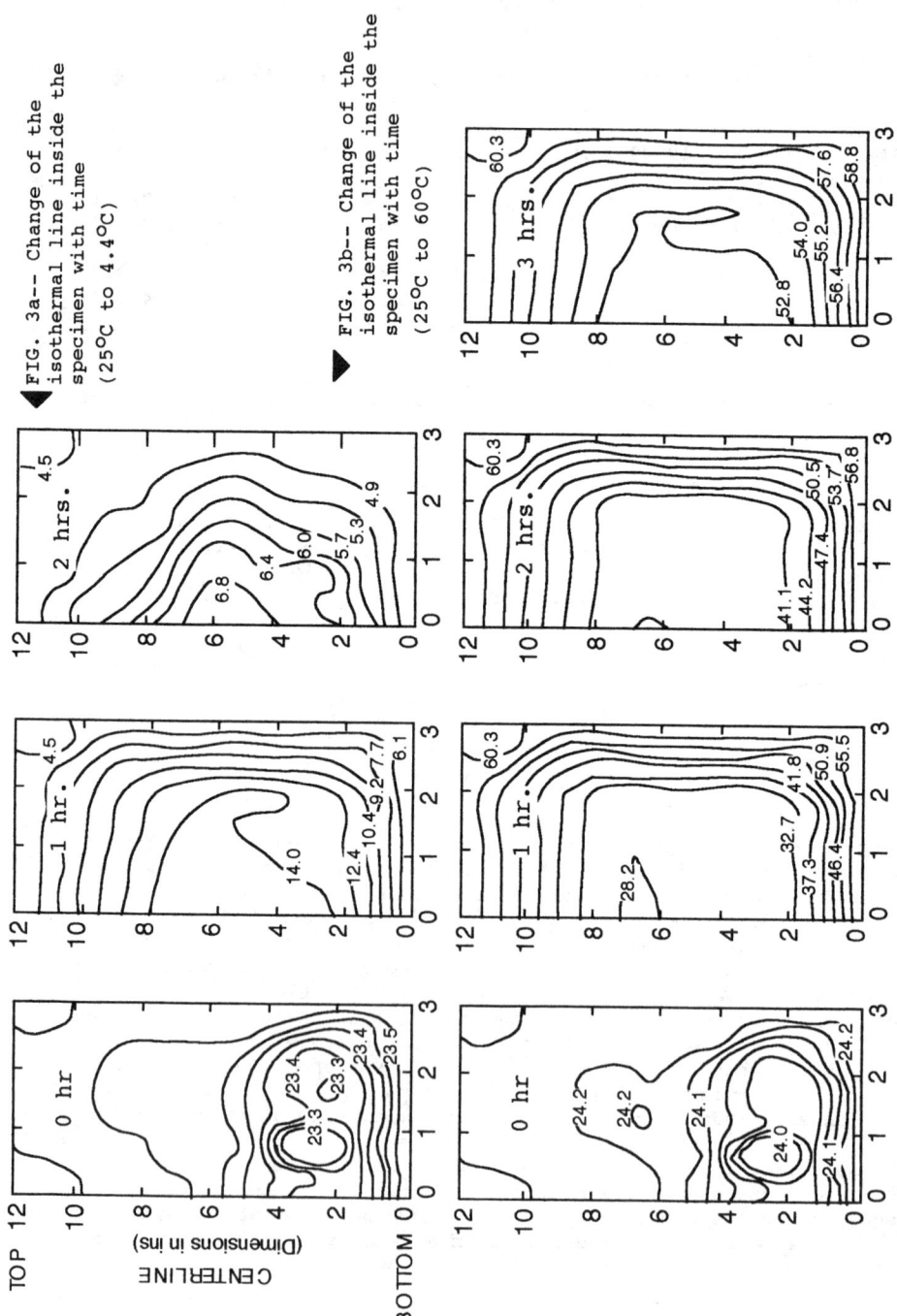

FIG. 3a-- Change of the isothermal line inside the specimen with time (25°C to 4.4°C)

FIG. 3b-- Change of the isothermal line inside the specimen with time (25°C to 60°C)

TEST RESULTS

All the test data were recorded using a data acquisition unit. FIG. 4 shows a typical set of strain- and stress-histories. The dynamic modulus, $|E^*|$, which is the absolute value of the complex modulus, is determined by the following equation.

$$|E^*| = \sigma_o / \varepsilon_o \tag{1}$$

where σ_o = axial loading stress, kPa, and
 ε_o = recoverable axial strain, in./in.

The values of σ_o and σ_o were obtained graphically from figures such as FIG. 4.

The phase angle can typically be obtained from the stress and the strain histories in the time domain in the manner as illustrated in FIG. 5. Alternatively, a simple way of determining the phase angle is in the frequency domain by applying the Fast Fourier Transform (FFT) [7] to the stress and the strain histories in the time domain and the stress and the strain spectra in the frequency domain is then obtained. The transfer function consists of each component of the displacement spectrum divided by the force spectrum at the corresponding frequency. Phase angle is the tangent of lag angle between the real and the imaginary component of transfer function (i.e., phase angle = imaginary component/real component) [8].

The summary of results obtained from the dynamic tests for four types of the asphalt concrete mixtures are shown in Table 4. The detailed results can be found elsewhere [9].

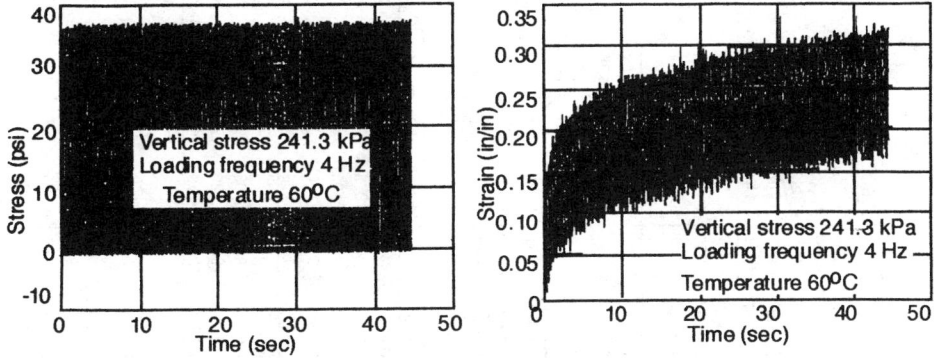

FIG. 4--Stress and strain histories

ANALYTICAL MODELING OF DYNAMIC MODULUS

The dynamic modulus is represented using a modified power law [10]. This representation is a simple generalization of power law and is capable of fitting data over a broad frequency range.

TABLE 4.—Magnitudes of dynamic modulus

(a) TYPE I-A

Axial Stress (kPa)	4.4 oC Hz	4.4 oC MPa	25 oC Hz	25 oC MPa	37.8 oC Hz	37.8 oC MPa	60 oC Hz	60 oC MPa
60.7	1.075	1186	1.056	910	1.075	704	1.075	531
	4.103	1351	4.083	952	4.083	869	4.103	724
	8.106	1455	8.067	1351	8.106	1131	8.126	862
	16.76	1565	16.7	1531	16.94	1510	16.8	986
121.3	1.075	1662	1.075	1310	1.075	731	1.075	545
	4.103	1751	4.083	1358	4.103	1041	4.103	827
	8.126	1917	8.087	1434	8.126	1248	8.126	958
	16.74	2179	16.76	1724	16.78	1503	16.76	1193
182							1.075	641
							4.103	855
							8.126	1062
							16.82	1234
241.3	1.075	2744			1.075	765		
	4.083	3296			4.103	1100		
					8.126	1510		
	16.76	4165			16.78	1613		

(b) TYPE I-B

Axial Stress (kPa)	4.4 oC Hz	4.4 oC MPa	25 oC Hz	25 oC MPa	37.8 oC Hz	37.8 oC MPa	60 oC Hz	60 oC MPa
60.7	1.075	1517	1.07	1276	1.07	896	1.07	607
	5.12	1813	4.14	1572	4.1	1055	4.1	869
	8.11	2427	8.11	1820	8.13	1303	8.15	1041
	16.8	2655	16.8	2262	16.8	1696	16.8	1310
121.3	1.075	2848	1.07	2620	1.07	1110	1.07	648
	4.1	2965	4.06	2620	4.1	1365	4.1	986
	8.13	3199	8.05	2958	8.11	1620	8.15	1131
	16.8	3372	10.7	3137	10.8	1738	16.8	1400
182							1.07	724
							4.1	986
							8.15	1214
							16.8	1476
241.3	4.1	4551	1.07	3310	1.09	1186		
	8.13	4937	4.1	3585	4.1	1607		
	16.8	5392	8.13	4165	8.11	1827		
			16.7	5075	16.8	2117		

(c) TYPE I-D

Axial Stress (kPa)	4.4 oC Hz	4.4 oC MPa	25 oC Hz	25 oC MPa	37.8 oC Hz	37.8 oC MPa	60 oC Hz	60 oC MPa
690.7	1.075	910	1.075	807	1.075	283	1.075	179
	4.103	1055	4.083	841	4.103	469	4.103	259
	8.126	1158	8.106	972	8.126	558	8.146	324
	16.8	1400	16.76	1179	16.78	676	16.82	359
121.3	1.075	1331	1.075	876	1.075	352	1.075	214
	4.103	1524	4.083	993	4.103	558	4.103	303
	8.126	1710	8.087	1007	8.126	738	8.126	400
	16.74	1855	16.76	1303	16.78	972	16.76	448
182							1.075	455
							4.103	662
							8.146	841
							16.82	979
241.3	1.075	1613	1.075	945				
	4.103	1717	4.083	1083	4.103	627		
	8.126	2117	8.104	1317	8.126	834		
	10.74	1986	16.69	1689	16.8	1089		

(d) ATOGB

Axial Stress (kPa)	4.4 oC Hz	4.4 oC MPa	25 oC Hz	25 oC MPa
60.7				
121.3	1.075	1083	1.075	565
	4.103	1303	4.103	1041
	8.146	1565	8.126	1193
	16.8	1576	16.8	1338

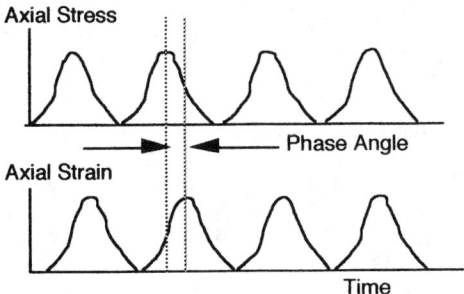

FIG. 5--Dynamic testing histories

The dynamic modulus modified power law is

$$|E^*|(f) = |E^*|_{min} + \frac{(|E^*|_{max} - |E^*|_{min})}{\left[1 + \left(\frac{h}{f}\right)\right]^m} \qquad (2)$$

where $|E^*|_{min}$ and $|E^*|_{max}$ are the minimum and maximum values of dynamic moduli with respect to frequency f, respectively; and, 'h' and 'm' are non-negative constants.

The constants 'm' is the value of the slope of the log $|E^*|$-log f plot in midway between the minimum and maximum values of the dynamic moduli; i.e.,

$$m = \frac{d \log |E^*|}{d \log f} \quad \text{at} \quad \left(|E^*| = |E^*|_{middle}\right) \qquad (3)$$

where

$$\log |E^*|_{middle} = \frac{1}{2}\left[\log |E^*|_{min} + \log |E^*|{max}\right] = \frac{1}{2} \log |E^*|_{min} |E^*|_{max} \qquad (4)$$

By taking the antilog, it is can be shown that modulus $|E^*|_{middle}$ is actually the geometric mean value

$$|E^*|_{middle} = \sqrt{|E^*|_{min} |E^*|_{max}} \qquad (5)$$

The frequency constant, h, can be calculated by matching the modified power law and the data at this same point; thus, by equating equation (2) with equation (5), it is found

$$h^{-1} = f_{middle}^{-1} \left[\left(\sqrt{\frac{|E^*|_{min}}{|E^*|_{max}}} + 1\right)^{1/m} - 1\right] \qquad (6)$$

where f_{middle} is the frequency at which the modulus data takes on the mean value described by equation (5).

The analytical representation of the master curve of dynamic moduli for three types of asphalt concrete mixtures is illustrated in FIG. 6. The measured data points are also shown in this figure. Table 5 shows the constants calculated for each of the analytical master curves.

TABLE 5--Constants used for construction of each master curve

Constants	Type I-A	Type I-B	Type I-D
$\|E^*\|_{min}$, MPa	537.8	620.6	241.3
$\|E^*\|_{max}$, MPa	2,151.2	3,130.3	1,565.2
m	0.485	0.56	0.667
h, Hz	0.085	0.135	0.156

From this figure it can be seen that the dynamic modulus of the asphalt concrete mixture Type I-D is more responsive to the loading frequency change than Type I-A or Type I-B.

Illustration: Using the Master Curve of the Dynamic Modulus

If it can be assumed that the diameter of contact area between tire and pavement surface is 12 in., and that the loading frequencies of vehicle travelling at 24 km/h and 104 km/h are 22 Hz and 95 Hz, respectively; the dynamic modulus of Type I-A increases from 1,654.8 MPa to 2,392.6 MPa.

ESTIMATION OF PERMANENT DEFORMATION

The dynamic tests were performed under controlled stress conditions. The test procedure of these tests was the same as those described in earlier sections. In laboratory repeated load testing, the resulting permanent strain levels are often reported with a corresponding number of repeated load levels. This is because it is recognized that the accumulation of residual deformation varies directly with the number of load repetitions. For the first few cycles of loading, the rate accumulation also varies significantly. However, after some number of load cycles, the rate of accumulation of residual deformation can be seen to reach a constant value. In the case of the specimens tested here, about 30 to 100 load-ing cycles were required to reach the level-off point.

It is assumed in this study that the logarithmic relationship between the number of repeated loads and permanent strain is essentially linear over a range of load applications. This is illustrated in FIG. 7. Experience of these authors had shown that the initial portion of the curve is very much dependent on the loading device and the manner in which the specimen is prepared. However, it appears that the slope subsequent portion of the curve prior to failure is often reproducible.

The line that is used to fit the straight portion of the curve in FIG. 7 is as follows.

$$\varepsilon^p = a N^b \tag{7}$$

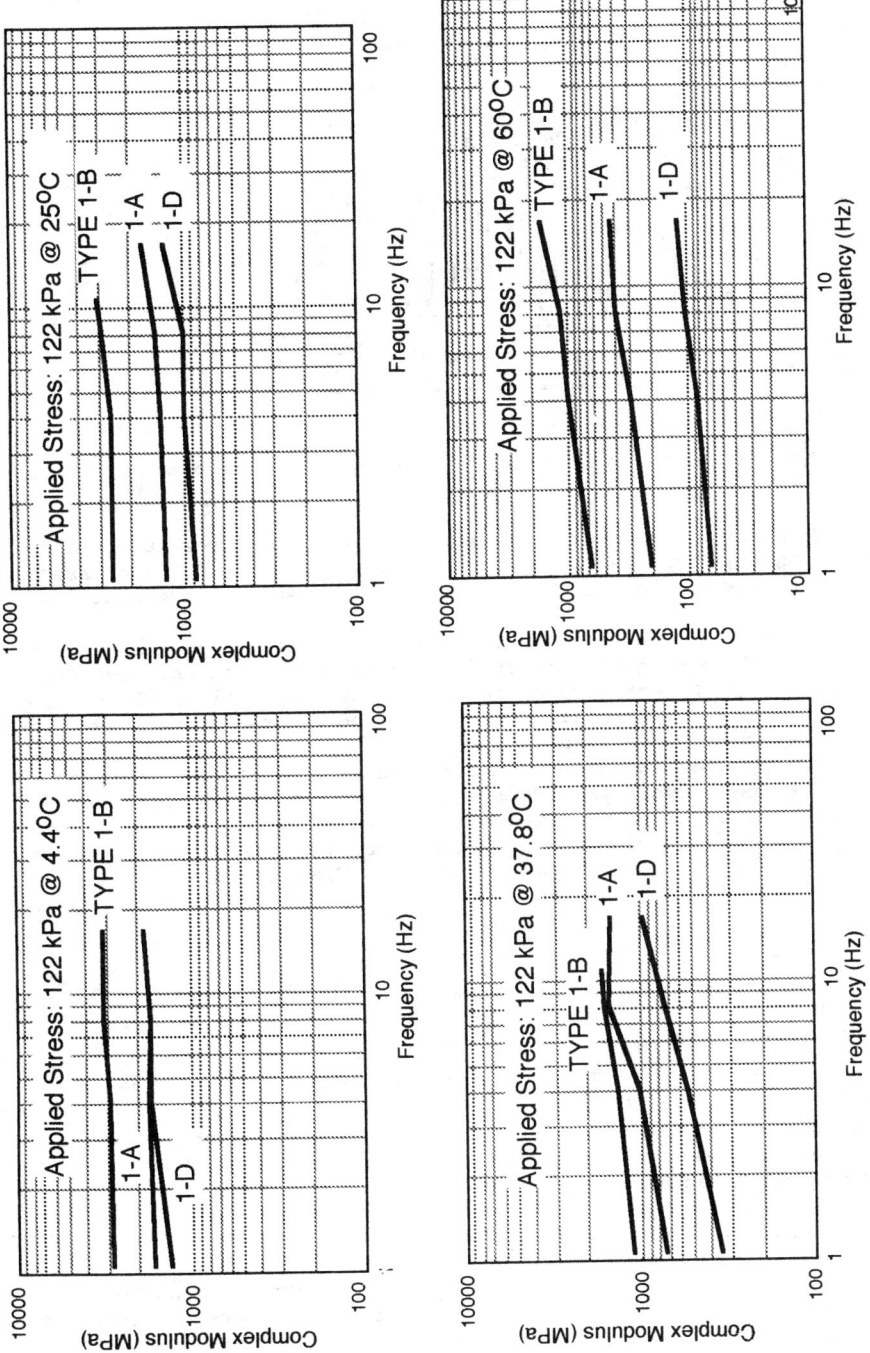

FIG. 6 -- Dynamic moduli of the asphalt mixtures at different frequencies (122 kPa)

where εp = total accumulated permanent strain at the end of N cycles,
a = intercept with permanent strain axis,
N = number of load applications, and
b = slope of the linear portion of the logarithmic relationship.

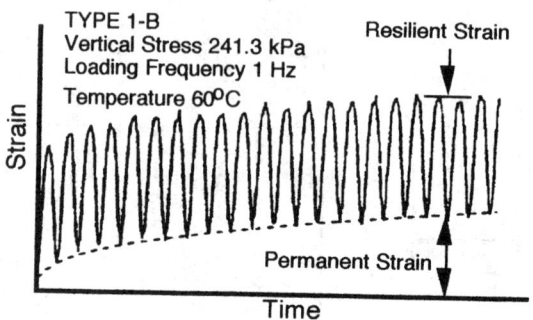

FIG. 7--Typical relationship between permanent strain and number of load repetitions

The influence of stress state, stress duration, and temperature are reflected in coefficients 'a' and 'b'. The coefficient 'a' increases with an increase in the loading rate but 'b' seems to go the opposite way.

Particular values of 'a' and 'b' can be described more accurately if the following are known: (1) stress distributions in the pavement system (which can be obtained using one of the many analytical equations [11]), (2) temperature profile with depth, and (3) variation of vertical stress pulse frequency with depth.

It is possible to estimate the permanent deformation occurring in a particular asphaltic layer. This is done by computing the permanent strain at a number of points within the layer, the number being sufficient to reasonably define the strain variation with depth. Summation of the permanent strain with depth under the wheel paths yields the permanent deformation at the surface, i.e.:

$$\delta_i^p = \sum_{n=1}^{n} \left(\varepsilon_i^p \Delta z_i \right) \tag{8}$$

where δp_i = rut depth in the ith position
εp_i = average permanent strain at depth $[z_i + (\Delta z_i/2)]$, and
Δz_i = difference in depth.

Clearly, because of the empirical nature of the prediction method care must be taken to ensure that the calculations do not go beyond the ranges of the variables investigated.

Illustration: Using the Rutting Data

Asphalt concrete surface layer of 10 cm thickness is assumed for each type of mixture. FIG. 8 shows variations of temperature, vertical stress, and frequency of vertical stress pulse with depth within the surface course. Vertical stress profile with depth is obtained according to Boussinesq's formula [11]. A single axle load of 80 kN and a tire pressure of 689.5 kPa are assumed. Variation of frequency of vertical stress pulse is constructed from Barksdale's results [12]. Vehicle speed of 72 km/h is considered in this illustration. From dynamic testing an empirical relations between magnitude of 'b' and loading frequency, temperature, and vertical stress are obtained for each type of mixture, i.e.:

$$b = 0.16016 + 0.001122\sigma_1 + 0.002617T - 0.01564F \text{ (Type I-A)}$$
$$b = 0.06767 + 0.000236\sigma_1 + 0.002351T - 0.00713F \text{ (Type I-B)} \quad (9)$$
$$b = 0.12977 + 0.000315\sigma_1 + 0.002524T - 0.01392F \text{ (Type I-D)}$$

where σ_1 = vertical stress (kPa),
T = temperature (°C), and
F = loading frequency (Hz).

For a value of 'a', average value is determined from results of dynamic testing. Values of 'a' = 0.0036, 0.0065, and 0.0063 for Type I-A, Type I-B, and Type I-D, respectively. Number of 80 kN single axle load [SAL] with 689.5 kPa tire pressure until 18.75 mm (3/4 in.) rut depth is presented in Table 6.

TABLE 6--Prediction of 80 kN SAL repetition until 18.75 mm rut

Mix Type	Number of 80 kN SAL
Type I-A	1,150,000
Type I-B	17,400,000
Type I-D	2,700,000

It can be seen here that Type I-B is much more resistant to rutting than Type I-A and I-B asphalt concrete.

PREDICTION OF FATIGUE CRACK PROPAGATION

The factors that affect the fatigue behavior of asphalt concrete in the field include the environment, stress state, load history and flaw type of the material. Paris [13] first suggested that for fatigue crack propagation in metal the crack growth rate was almost exclusively dependent on the change in the stress intensity factor during each cycle (ΔK). The amplitude of the stress field near the crack tip is referred to as the "Stress Intensity Factor", K. At the point in which the crack is about to propagate, this value will be equal to that of the critical stress intensity factor or K_{IC}, which is also often referred to as the fracture toughness of the material. The Paris equation which describes the crack growth rate is

$$\frac{dc}{dN} = A(\Delta K)^n \quad (10)$$

where c = crack length,

N = number of cycles,
ΔK = stress intensity amplitude ($\Delta K = K_{max} - K_{min}$), and
A, n = material constants obtained from experimental tests.
The prediction of total fatigue life, N_f can be obtained by integrating equation (10) and the number of cycles to failure is

$$N_f = \int_{c_o}^{c_f} \frac{1}{A(\Delta K)^n} dc \qquad (11)$$

where c_o = initial crack size, and
c_f = final crack size.

Schapery [14] developed a general theory of crack growth in viscoelastic media. This theory of fatigue crack propagation in asphaltic concrete was derived from the power law relation of equation (10), and according to Schapery,

$$\frac{dc}{dN} = B_t (\Delta K)^{2\left(1+\frac{1}{m}\right)} \qquad (12)$$

where

$$B_t = \frac{(1-\nu^2) D_1 \lambda_m^{1/m} \Pi^m}{2^{m+1}} \int_t^{t+t_p} \frac{W^2 \left(1+\frac{1}{m}\right)}{\gamma^{1/m} \sigma_m^2 I_1^2} dt \qquad (13)$$

m = slope of the straight line region of the creep compliance curve,

$$\lambda_m = \frac{3\left(\Pi^{1/2}\right)\Gamma(m+1)}{4\left(m+\frac{3}{2}\right)\Gamma\left(m+\frac{3}{2}\right)}, \quad \Gamma(m) = \int_0^\infty t^{m-1} e^{-t} dt \qquad (14)$$

D_1 = intercept of straight line with log t = 0 in the creep compliance curve,
$\Gamma(m)$ = the Gamma function with argument 'm',
W = W(t) which defines the wave shape of the stress intensity factor,
γ = fracture energy density,
σ_m = maximum stress a material can sustain,
I_1 = integral measure of the shape of the stress distribution in the failing material ($0 < I_1 \le 2$, with $I_1 = 2$ when this stress is uniformly distributed), and
t_p = time of loading for a given stress pulse.

The parameter $\lambda_m^{1/m}$ depends only on 'm' ($\lambda_m^{1/m} \approx 1/3$ for $0 \le m \le 1$). If it can be assumed that (1) asphalt concrete mixture is a linearly viscoelastic material, and (2) the extended correspondence principle established by Graham [15] is applicable, then equality of the empirical equation (10) and the theoretical equation (12) requires that:

$A = B_t$, and $n = 2(1+1/m)$.

The problem of a growing crack in viscoelastic medium or that of a crack of constant length but subjected to a time variation in loading cannot, in general, be treated by the correspondence principle. For one important case, however, when the loading increases monotonically with time, Graham has shown that the distribution of stresses and strains around a crack can be found by an extended correspondence principle [16]. This result enables the viscoelastic response to be predicted and then related to the fatigue constants.

Illustration: Using the Fatigue Data

Prediction of fatigue life of three types of asphalt concrete mixtures was made. The following calculations must be made in order to obtain B_t. Firstly, the quantity of fracture energy, γ, is obtained by the following relationship:

$$\gamma = \frac{C_e K_I^2}{8} \tag{15}$$

where $C_e = 4(1-v^2)/|E^*|$ (v = Poisson's ratio), and

K_I = stress intensity factor for the opening mode.

Secondly, the magnitude of σ_m can be obtained from the result of the Marshall test which is a routine test for asphalt concrete mixture design.

$$\sigma_m = \frac{2P}{\pi t d} \tag{16}$$

where P = Marshall stability,
t = specimen thickness, and
d = specimen diameter.

The value of D_1 can be derived from the following equation [17]:

$$D_1 = \frac{1}{|E^*|_{min}} \frac{\sin(m\pi)}{m\pi} \tag{17}$$

Assume the Poisson's ratio for the asphaltic road material to be 0.4. Assume that the wave shape of stress intensity factor, $W(t)$, is a haversine function and the vehicle is traveling at 48 km/h over this material. Next, determine the stress intensity factor, K_I, from FIG. 9 which is developed from finite element approach [18] to predict a reflection cracking in overlay. Modulus ratios, E_1/E_2 (stiffness of upper layer/stiffness of underlying layer) of 1, 2, and 3 are used in this example.

The fatigue life, N_f, is then obtained by the simplification of equation (11) such that

$$N_f = \frac{h}{A K_I^n} \tag{18}$$

where h = thickness of surface asphalt concrete layer.

228 ENGINEERING PROPERTIES OF ASPHALT MIXTURES

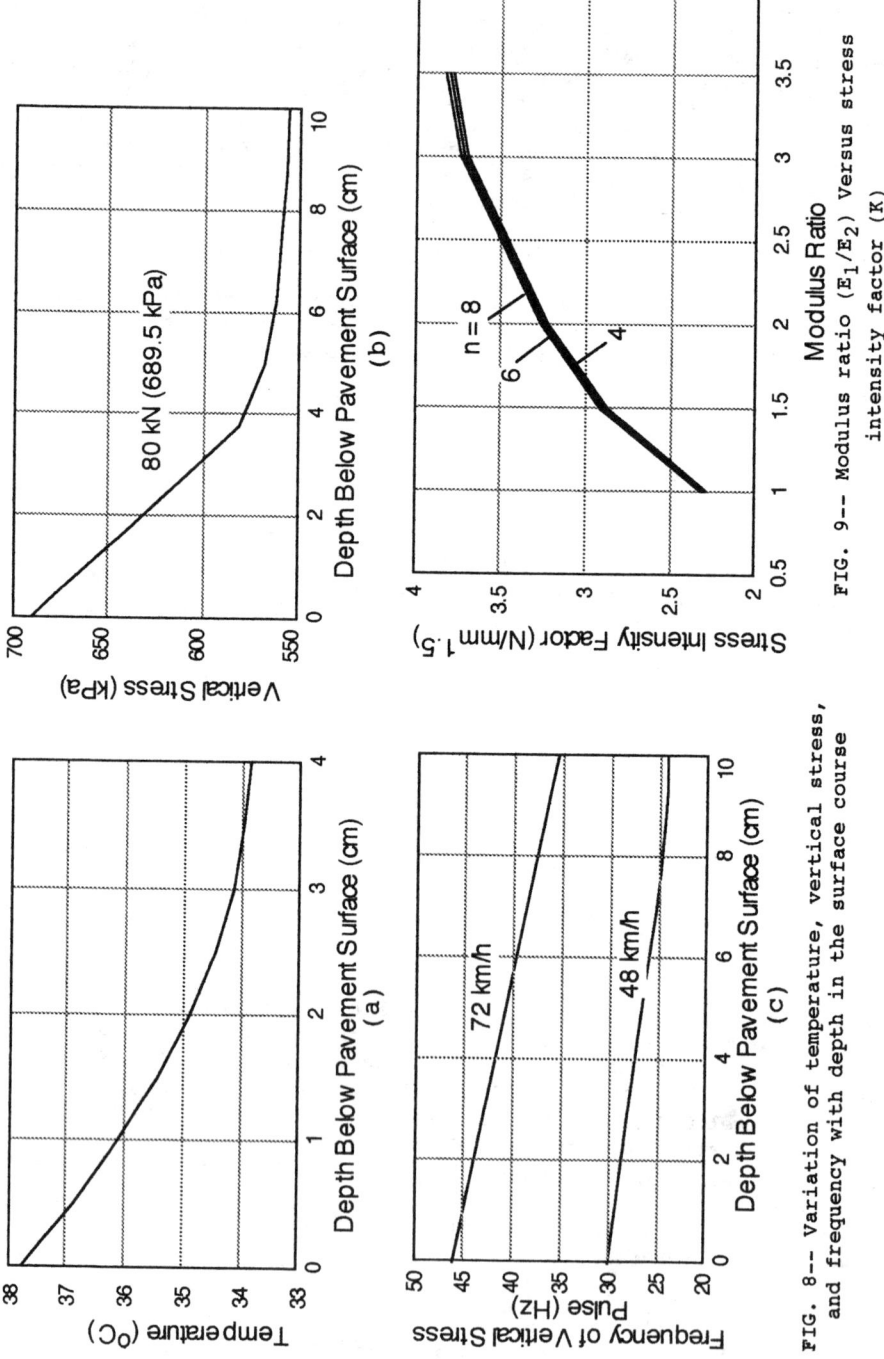

FIG. 8 -- Variation of temperature, vertical stress, and frequency with depth in the surface course

FIG. 9 -- Modulus ratio (E_1/E_2) versus stress intensity factor (K)

$A = B_t$, and $n = 2(1+1/m)$.

The problem of a growing crack in viscoelastic medium or that of a crack of constant length but subjected to a time variation in loading cannot, in general, be treated by the correspondence principle. For one important case, however, when the loading increases monotonically with time, Graham has shown that the distribution of stresses and strains around a crack can be found by an extended correspondence principle [16]. This result enables the viscoelastic response to be predicted and then related to the fatigue constants.

Illustration: Using the Fatigue Data

Prediction of fatigue life of three types of asphalt concrete mixtures was made. The following calculations must be made in order to obtain B_t. Firstly, the quantity of fracture energy, γ, is obtained by the following relationship:

$$\gamma = \frac{C_e K_I^2}{8} \tag{15}$$

where $C_e = 4(1-v^2)/|E^*|$ (v = Poisson's ratio), and

K_I = stress intensity factor for the opening mode.

Secondly, the magnitude of σ_m can be obtained from the result of the Marshall test which is a routine test for asphalt concrete mixture design.

$$\sigma_m = \frac{2P}{\pi t d} \tag{16}$$

where P = Marshall stability,
t = specimen thickness, and
d = specimen diameter.

The value of D_1 can be derived from the following equation [17]:

$$D_1 = \frac{1}{|E^*|_{min}} \frac{\sin(m\pi)}{m\pi} \tag{17}$$

Assume the Poisson's ratio for the asphaltic road material to be 0.4. Assume that the wave shape of stress intensity factor, $W(t)$, is a haversine function and the vehicle is traveling at 48 km/h over this material. Next, determine the stress intensity factor, K_I, from FIG. 9 which is developed from finite element approach [18] to predict a reflection cracking in overlay. Modulus ratios, E_1/E_2 (stiffness of upper layer/stiffness of underlying layer) of 1, 2, and 3 are used in this example.

The fatigue life, N_f, is then obtained by the simplification of equation (11) such that

$$N_f = \frac{h}{A K_I^n} \tag{18}$$

where h = thickness of surface asphalt concrete layer.

228 ENGINEERING PROPERTIES OF ASPHALT MIXTURES

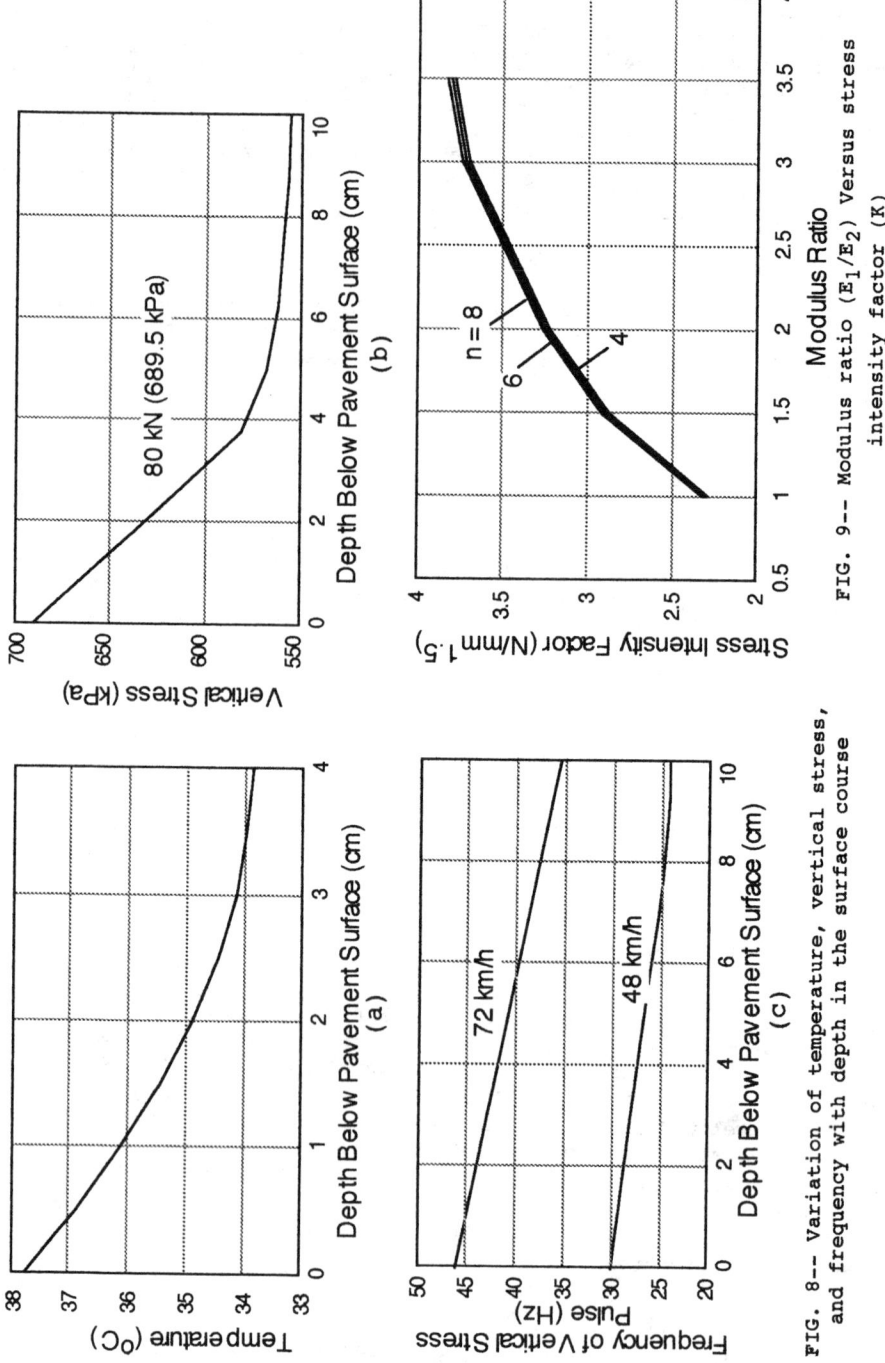

FIG. 8 -- Variation of temperature, vertical stress, and frequency with depth in the surface course

FIG. 9 -- Modulus ratio (E_1/E_2) versus stress intensity factor (K)

If it can be further assumed that the layer thickness, h, is 7.5 cm, and using the laboratory determined fatigue constants (A, n), the fatigue life for the different asphalt concrete mixtures studied can be obtained. Table 7 shows the predictions of the number of load cycles, N_f, which will cause cracking.

TABLE 7--Results of fatigue life prediction

Mix type	A (mm/cycle)	n	N_f (cycles)		
			$E_1/E_2=1$	$E_1/E_2=2$	$E_1/E_2=3$
Type I-A	3.35E-7	6.12	1,368,000	165,000	72,000
Type I-B	4.6E-8	5.57	15,757,000	2,297,000	1,082,000
Type I-D	1.55E-6	5.0	752,000	133,000	68,000

From this illustration it can be seen that a higher modular ratio between overlay material and existing pavement will increase the stress intensity factor and result faster propagation of reflection cracking. It can be seen that Type I-B is also superior to Type I-A and I-D in terms of fatigue.

CONCLUDING REMARKS

A comprehensive characterization of the performance-related properties of the four types of an asphalt concrete mixtures used by NMSH&TD is presented here. The performance-related properties include rutting and fatigue behavior and were obtained by dynamic modulus testing. It appears from the performance-related characteristics of asphalt concrete mixtures that Type I-B is in many ways superior to Type I-A and I-D. The performance of last two appears to be only marginally different. It may be possible then for highway engineers to avoid the more involved and costly procedures to obtain similar results.

ACKNOWLEDGMENTS

This work was supported by the New Mexico State Highway and Transportation Department. The authors are very grateful to Mr. Dick Leuck and Mr. David Catanach of the Preliminary Design Bureau, and Mr. Jim Stokes of the Materials Laboratory, for their support and interest in this study.

REFERENCES

[1] AASHTO guide for design of pavement structures, Amer. Assoc. of State Highway and Transp. Officials, Washington, DC, 1986.

[2] Matthews, J.M., Monismith, C.L., and Craus, J., "Investigation of Laboratory Fatigue Testing Procedures for Asphalt Aggregate Mixtures," J. of Transp. Engrg., Vol.119, No.4, pp. 634-654, July/August, 1993.

[3] Supplemental Specifications Section 401-Plant Mix Bituminous Pavement (PMBP), New Mexico State Highway and Transportation Dept., 1992.

[4] Papazian, H.S. and Baker, R.G., "Analysis of Fatigue Type Properties of Bituminous Concrete," Proceedings, AAPT, Vol. 28, 1959.

[5] Shook, J.F., and Kallas, B.F., "Determining Material Properties," Paper Presented at Institute on Flexible Pavement Design and Performance, Pennsylvania State Univ., University Park, Pa., Nov. 12-16, 1973.

[6] Bautista, M.A.V., "Placement Conditions and Detection of Moisture Movement in Crushed Rock Salt Backfill," Thesis, Department of Civil Engineering, Univ. of New Mexico, 1993.

[7] Cooley, J.W., and Tukey, J.W., "An Algorithm for the Machine Calculation of Complex Fourier Series," Mathematics of Computation, Vol. 19, No. 90, 1965, pp. 297-301.

[8] Chua, K.M. and Lytton, R.L., "Dynamic Analysis Using the Portable Pavement Dynamic Cone Penetrometer," Transportation Research record 1192, TRB, National Research Council, Washington, D.C., pp. 27-38, 1989.

[9] Roo, M.C., "Comprehensive Characterization of Asphalt Concrete Mixtures and Pavement Soil Materials," Dissertation, Department of Civil Engineering, Univ. of New Mexico, 1994.

[10] Schapery, R.A., Unpublished Class Note at Texas A&M Univ., 1983.

[11] Yoder, E.J. and Witczak, M.W., "Principles of Pavement Design," 2nd Edition, John Wiley & Sons, Inc., pp. 24-80, 1975.

[12] Barksdale, R.D., "Compressive Stress Pulse Times in Frexible Pavements for Use in Dynamic Testing," Highway Research Record 345, Highway Research Board, pp. 32-44, 1971.

[13] Paris, P.C. in proceedings, 10th Sagamore Army Materials Research Conference, Syracuse University Press, 1964.

[14] Schapery, R.A., "A Theory of Crack Growth in Viscoelastic Media," Texas A&M University, Report MM 2764-73-1, March 1973.

[15] Graham, G.A.C., "The Correspondence Principle of Linear Viscoelasticity Theory for Mixed Boundary Value Problems Involving Time-Dependent Boundary Regions, Q. Applied Math., Vol.26, pp.167-174, 1968.

[16] Wnuk, M.P., and Knauss, W.G., "Delayed Fracture in Viscoelastic-Plastic Solids," Int. J. Solids Structures, Vol.6, pp.995-1009, 1970.

[17] Pipkin, A.C., "Lectures on Viscoelasticity Theory," 2nd Edition, Applied Mathematical Sciences Vol. 7, Springer-Verlag, pp. 22-32, 1986.

[18] Brooker, T., Foulkes, M.D., and Kennedy, C.K., "Influence of Mix Design on Reflection Cracking Growth Rates Through Asphalt Surfacing," Proceedings of 6th International Conference on the Structural Design of Asphalt Pavements, pp. 107-120, Univ. of Michigan, Ann Arbor, Michigan, July 17-19, 1987.